MOLECULAR
BIOLOGY
INTELLIGENCE
UNIT

UNUSUAL SECRETORY PATHWAYS: FROM BACTERIA TO MAN

Karl Kuchler

Department of Molecular Genetics
University and Biocenter of Vienna
Vienna, Austria

Anna Rubartelli

Istituto Nazionale per la
Ricerca sul Cancro
Servizio di Patologia Clinica
Genova, Italy

Barry Holland

Institut de Génétique et Microbiologie
Université Paris XI
Orsay, France

Springer-Verlag Berlin Heidelberg GmbH

MOLECULAR BIOLOGY INTELLIGENCE UNIT
UNUSUAL SECRETORY PATHWAYS: FROM BACTERIA TO MAN

R.G. LANDES COMPANY
Austin, Texas, U.S.A.

International Copyright © 1997 Springer-Verlag Berlin Heidelberg
Originally published by Springer-Verlag in 1997
Softcover reprint of the hardcover 1st edition 1997

 Springer

ISBN 978-3-662-22583-7   ISBN 978-3-662-22581-3 (eBook)
DOI 10.1007/978-3-662-22581-3

While the authors, editors and publisher believe that drug selection and dosage and the specifications and usage of equipment and devices, as set forth in this book, are in accord with current recommendations and practice at the time of publication, they make no warranty, expressed or implied, with respect to material described in this book. In view of the ongoing research, equipment development, changes in governmental regulations and the rapid accumulation of information relating to the biomedical sciences, the reader is urged to carefully review and evaluate the information provided herein.

Library of Congress Cataloging-in-Publication Data
Catalog number applied for, but not available at press time.

PUBLISHER'S NOTE

R.G. Landes Bioscience Publishers produces books in six Intelligence Unit series: *Medical, Molecular Biology, Neuroscience, Tissue Engineering, Biotechnology* and *Environmental*. The authors of our books are acknowledged leaders in their fields. Topics are unique; almost without exception, no similar books exist on these topics.

Our goal is to publish books in important and rapidly changing areas of bioscience for sophisticated researchers and clinicians. To achieve this goal, we have accelerated our publishing program to conform to the fast pace at which information grows in bioscience. Most of our books are published within 90 to 120 days of receipt of the manuscript. We would like to thank our readers for their continuing interest and welcome any comments or suggestions they may have for future books.

Shyamali Ghosh
Publications Director
R.G. Landes Company

CONTENTS

EDITORS

Karl Kuchler
Department of Molecular Genetics
University and Biocenter of Vienna
Vienna, Austria
chapter 2

Anna Rubartelli
Istituto Nazionale per la
Ricerca sul Cancro
Servizio di Patologia Clinica
Genova, Italy
chapter 3

Barry Holland
Institut de Génétique et Microbiologie
Université Paris XI
Orsay, France
chapter 1

CONTRIBUTORS

Mark A. Blight
Institut de Génétique
et Microbiologie
Université Paris XI
Orsay, France
chapter 1

Christian Chervaux
Institut de Génétique
et Microbiologie
Université Paris XI
Orsay, France
chapter 1

Giovanna Chimini
Centre d'Immunologie
INSERM-CNRS
de Marseille Luminy
chapter 5

Ralf Egner
Department of Molecular Genetics
University and Biocenter
of Vienna
Vienna, Austria
chapter 2

Hans Geuze
Department of Cell Biology
Institute for Biomembranes
Utrecht University
Utrecht, The Netherlands
chapter 6

Yannick Hamon
Centre d'Immunologie
INSERM-CNRS
de Marseille Luminy
chapter 5

Marie Françoise Luciani
Centre d'Immunologie
INSERM-CNRS
  de Marseille Luminy
chapter 5

Kurt Pawlitschko
Max-Planck-Institut
  für Biochemie
Martinsried, Germany
chapter 4

Andréa de Lima Pimenta
Institut de Génétique et
  Microbiologie
Université Paris XI
Orsay, France
chapter 1

Graça Raposo
Institut Curie
Section de Recherche
Paris, France
chapter 6

Roberto Sitia
Immunologia Molecolare
DIBIT-HSR
Milan, Italy
chapter 3

Robert Tampé
Max-Planck-Institut
  für Biochemie
Martinsried, Germany
chapter 4

Stephan Uebel
Max-Planck-Institut
  für Biochemie
Martinsried, Germany
chapter 4

Stefanie Urlinger
Max-Planck-Institut
  für Biochemie
Martinsried, Germany
chapter 4

Michel Vidal
Univ. Montpellier II
Montpellier, France
chapter 6

PREFACE

Cell growth, differentiation, development and inter- and intra-cellular communication of all living cells or organisms depend upon highly coordinated secretory processes. The classical, N-terminal signal sequence-dependent secretory pathway accounts for constitutive protein export or in response to certain extracellular stimuli, for regulated secretion of polypeptides through secretory granules. However, it is increasingly appreciated and recognized that mechanisms for protein secretion must exist that operate independently and/or in parallel to the usual classical secretory pathways, as they do not require the presence of typical hydrophobic signal peptides or involve classical storage granules. Notably, many of these processes have major implications in the pathology of human disease, including auto-immune and infectious disease, cancer and hereditary abnormalities.

Thus, this book will provide a comprehensive discussion of currently known mechanisms of unusual routes for protein secretion as described from bacteria to man. Secretion of bacterial proteases and toxins such as hemolysin will be covered in detail. Peptide pheromone secretion in lower eukaryotes such as yeast will also be reviewed. In addition, we shall deal with the function of the mammalian peptide transporters required for antigen presentation. Facts and myths about possible release mechanisms for mammalian cytokines such as interleukin-1 and the fibroblast growth factors will be extensively discussed. Moreover, novel vesicular transport pathways in cells of the immune system such as those involved in the generation of cell-surface targeted exosomes will be discussed in detail. Finally, by analogy to bacterial and yeast ATP binding cassette transporters involved in unconventional secretory pathways, individual chapters of this book are devoted to discuss a hypothesized function of mammalian ABC transporters in the secretion of polypeptides and proteins via *hitherto* unknown secretory pathways.

========= CHAPTER 1 =========

PROTEIN SECRETION IN GRAM-NEGATIVE BACTERIA

Andréa de Lima Pimenta, Mark A. Blight,
Christian Chervaux and I. Barry Holland

I. INTRODUCTION

Since the initial characterization of the hemolysin secretion system (Hly) from the Gram-negative bacterium, *E. coli*, by W. Goebel's group in the late 1970s, there has been a surprising proliferation of discoveries of distinct protein secretion mechanisms in many Gram-negative bacteria. Both for the Hly system and for other secretion pathways, developments have been the most dramatic over the last 5 years and, therefore, in this review we have placed most detailed emphasis upon this period up to approximately mid-1996. Previous reviews have generally agreed on the classification of secretion Types I, II, III, indicated in this review. Here (see also reference 5), we have proposed a classification for additional pathways: Type IV (auto-transporter systems like the IgA pathway); Type V for surface pilins (*E. coli* Pap system), the functionally related Type Va, although not homologous with the pilin system, are transported from the periplasm via a single outer membrane "translocator" protein; Type VI for the special case of the filamentous phage (nucleo-protein) secretory pathway. The secretion of flagellar proteins on to the cell surface, although not discussed in this review, could be considered yet another pathway (for review see reference 5a). However, the biogenesis of the flagellum involves at least eight proteins with homology to proteins of the Type III pathway. We emphasize that Types IV-VI, as defined here, are useful working classifications but a generally agreed classification for these secretion pathways in the literature has not yet emerged. What can be generally agreed upon is that Types I, III and VI are one step processes with translocation from cytoplasm directly on to the cell surface or to the medium, while other pathways employ a two-step mechanism involving initial export to the periplasm targeted by an N-terminal signal via

Unusual Secretory Pathways: From Bacteria to Man, edited by Karl Kuchler, Anna Rubartelli and Barry Holland. © 1997 R.G. Landes Company.

the Sec-machinery (general export pathway). The second stage then involves translocation across the outer membrane by various mechanisms. Importantly, the proteins of the latter group are apparently translocated across the outer membrane in a fully folded form, while proteins translocated directly to the outside of the cell (Types I, III) and perhaps IV may be required to re-fold on the surface.

In the case of the one-step secretion pathways, Type I (HlyA) targeting to the secretion machinery or translocator complex involves a novel C-terminal (noncleaved) secretion signal, while Type III (e.g., Yop proteins) employs a novel N-terminal signal, again apparently uncleaved. As indicated above, the two-step pathways involve initial targeting, by classical N-terminal, processed secretion signals, to the periplasm via the Sec-machinery. Subsequent targeting to the outer membrane "translocator," which in many cases may be a relatively well conserved homo-multimeric channel, appears to involve one or two relatively large blocks of amino acids well separated in the molecule for at least one Type II protein (pullulanase).

The composition of the transenvelope Type I translocator, including an ATP-dependent ABC transporter, an additional inner membrane protein of the MFP family(membrane fusion protein) which spans the periplasm, and an outer membrane protein forming the presumed exit, is now well established. Nevertheless, the precise mechanism of translocation is largely unknown and studies have so far been restricted to in vivo analyses.

In the case of Type III secretion, also involving a single secretion step directly to the medium, some proteins of the presumed transenvelope complex have also been identified. More proteins than for the Type I translocator appear to be required and, surprisingly, so far there is no indication of any functional overlaps with the Type I system. In contrast, and a recurring theme in this review, Type III export across the outer membrane does involve shared homologs with other protein transport pathways, for example, with at least eight proteins involved in flagellar biogenesis. Similarly, the Type II pathway, exemplified by pullulanase secretion, requires proteins homologous to those involved in the biogenesis of *Pseudomonas* (non-Pap) pili and in filamentous phage secretion (Type VI). Several pathways are, therefore, composed of unique elements, in addition to elements with shared homologies with one or more different pathways. Moreover, several pathways (Types II, IV, V) use an identical (Sec) machinery for first-stage transport to the periplasm. In remarkable contrast, prepro-proteins of the Type IV pathway, following initial transport into the periplasm, engineer their own secretion, without apparent ancilliary proteins, by autotransport through a channel of their own making, followed by auto-proteolytic release of a secreted fragment.

In Gram-negative bacteria it clearly now transpires that protein secretion is a very widespread phenomenon with frequently even the same species solving the problem of transport to the exterior across the double envelope membrane, by different mechanisms. As a climax to the flood of recent discoveries, we are now coming to the end of the initial, relatively descriptive phase of these novel secretion pathways and two major questions are emerging—how do proteins utilizing the two-step pathway, first fold in the periplasm and then somehow traverse the outer membrane, while proteins directly secreted from cytoplasm to the exterior fold on the cell surface? The fundamental question—relevant to all protein transport mechanisms as to precisely how these proteins cross the cytoplasmic membrane—can also be addressed with these transport systems which offer the advantage, with respect to genetic manipulations, that they are nonessential. In many cases, the corresponding translocator proteins have been identified and targeting signals in different protein substrates characterized. Moreover, overproduction and structural analysis of some key translocator proteins is underway and should soon be realizing major new findings. A limiting factor is however at this stage the absence of any in vitro transport system and this must now be seriously addressed especially for the less complex systems. Finally, we are happy to draw attention to the fact that many of these secretion pathways are implicated in major pathogenicity mechanisms both in plants and animals and one can hope that this will provide the financial motor and the academic stimulation to solve outstanding technical problems to access the detailed mechanism of these novel secretion processes in the near future.

In Gram-negative bacteria proteins having an extra-cytoplasmic destination can be targeted to four different "compartments" within or outside the cell: the inner membrane, the periplasmic space, the outer membrane or the external medium. Transport of proteins targeted to the periplasm or to the outer membrane takes place in most cases through an apparently universal mechanism, the General Export Pathway (GEP).[1] This is primarily represented by the *Sec* machinery, which translocates proteins across the inner membrane, containing a specific N-terminal signal secretion sequence (Fig. 1.1).

However, in the case of the complete secretion of proteins to the medium through both membranes of Gram-negative bacteria, it appears, as summarized in Figure 1.1, that several different mechanisms have been evolved to deliver the allocrite (the "substrate of secretion")[2] to the external medium. At least six of these secretion pathways have been described so far,[3-5] designated "secretion routes I to VI" which includes the mechanism of transport of filamentous phage nucleo-protein particles (Table 1.1). These can generally be divided into two major sub-groups: (i) those in which secretion to the medium takes place in two steps, with a periplasmic intermediate (routes II, IV and V); and (ii)

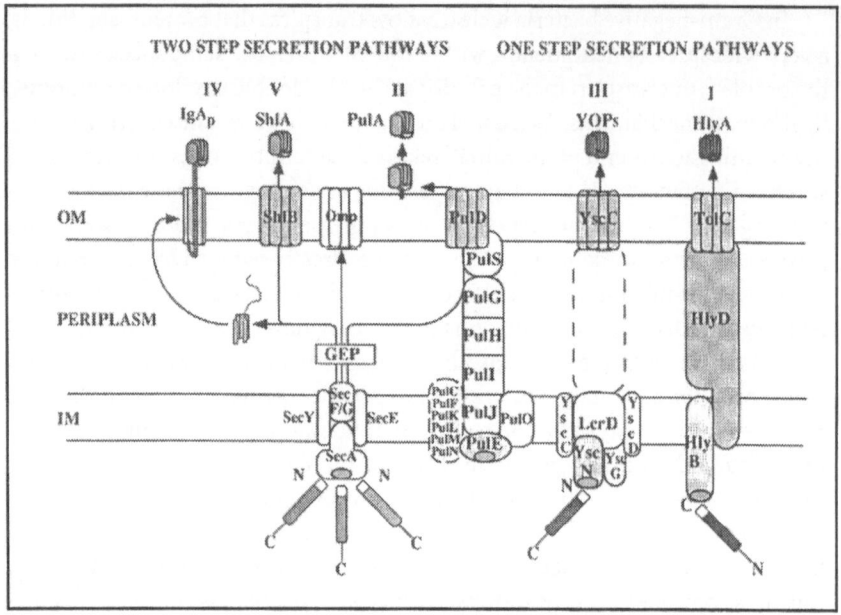

Fig. 1.1. Secretion pathways used by Gram-negative bacteria to deliver proteins to the extracellular medium (I, II, III, IV and V). GEP = General Export Pathway, IgA_p = N. gonorrhoeae IgA protease; Shl = S. marcescens non-RTX hemolysin; Pul = K. oxitoca pullulanase; Ysc = Y. pestis YOP proteins; Hly = E. coli hemolysin. Pathway VI, responsible for filamentous phage secretion, was excluded from the diagram for simplicity (but see text for details). ATP binding proteins are indicated by a shaded elipse, e.g., SecA.

those delivering proteins directly to the external medium in one single step (routes I, III and VI). In the cases where secretion occurs in a two-step manner, the first step, i.e., translocation across the cytoplasmic membrane, takes place through the GEP. Other more dedicated mechanisms then recruit these periplasmic intermediates to accomplish the final step of their secretion through the outer membrane to the external medium. Secretion directly to the medium, on the other hand, takes place through GEP-independent routes which bypass the periplasm and are totally specific to each allocrite secreted.

II. TWO-STEP SECRETION PATHWAYS: THE GSP

As shown in Figure 1.1, three different secretion systems use the GEP as a route to cross the inner membrane and deliver their allocrites into the periplasmic space, from where they are finally translocated across the outer membrane to the external medium. These three mechanisms, which may either be considered individually as different secretory pathways or as different terminal branches of what has been called the *General Secretory Pathway* (GSP;

Table 1.1. The six secretion pathways used by Gram-negative bacteria

Secretion Pathway	Organism[a]	Allocrite[b]	Secretion Signal[c]	Steps	Export GEP/Sec dependent	Periplasmic Intermediate	Accessory proteins for secretion
I	Escherichia coli	HlyA (RTX)	C-terminal	1	NO	NO	HlyB (IM), HlyD (IM), TolC (OM)
II	Klebsiella oxytoca	PulA	? (N-terminal[d])	2	YES	YES-Folded	At least 14 proteins
III	Yersinia pestis	Yops	N-terminal	1	NO	?	At least 15 proteins
IV	Neisseria gonorrhoeae	IgA	Auto-transport (C-terminal domain)	2	YES	YES-Unfolded	Auto-transport
Va	Escherichia coli	Pilins	N-terminal	2	YES	YES-Folded	PapC (OM); PapD chaperone (PE)
Vb	Serratia Marcescens	Sh1A	N-terminal	2	YES	YES	Sh1B (OM)
VI	filamentous phages	phage particle	?	1	NO	NO	pI (IM) and PpIV (OM)

[a]Specific examples are indicated here, other organisms are mentioned in the text; [b]substrate secreted through each system by the organisms mentioned in [a]; IM = Inner Membrane; PE = Periplasm; OM = Outer Membrane; [c]secretion signal responsible for translocation across the outer membrane, as opposed to the classical N-terminal Sec signal-peptide; [d]identified between the residues 60-120, in the case of P. aeruginosae exotoxin A.[35]

see reference 6) are represented by: (i) secretion pathway II, which uses the products of at least 14 genes to translocate the allocrite across the outer membrane; (ii) pathway IV, in which the last step of secretion depends only on the allocrite itself and (iii) pathway V, in which translocation of the allocrite from the periplasm to the external medium is accomplished by the products of one or two genes, one being a periplasmic chaperone. These three pathways will now be considered in some detail before describing the one-step routes, III, VI. Finally, Type I secretion, the pathway most extensively studied so far, will be presented in the last part of this review.

SECRETION PATHWAY II: THE PUL SYSTEM

This is the terminal branch of the GSP used by many Gram-negative bacteria to deliver proteins into the external medium. Examples of secretion systems included in this category are: the Pul systems of *Klebsiella oxytoca*[7] and *K. pneumoniae*;[8] the Out systems of *Erwinia chrysanthemi*[9] and *E. caratovora*;[10] the Xps system of *Xanthomonas campestris*;[11] the Exe system of *Aeromonas hydrophila*[12-14] and the Xcp or Pil system of *Pseudomonas aeruginosa*.[12] It is important to note here that homologous proteins of the Type II secretion pathway can be involved in the assembly of surface pili (i.e., Type IV pili as found in *Pseudomonas*), DNA transfer mechanisms and filamentous phage release as well as protein secretion in different organisms.[15,16] The best studied Type II system is the Pul system of *K. oxytoca*, which is responsible for the secretion of the enzyme pullulanase (PulA), and the details of this pathway and proteins involved are illustrated in Figure 1.2.

Pullulanare, PulA, is a 117 kDa lipoprotein which, together with the proteins required for secretion, is encoded by a cluster of 15 genes organized as two adjacent operons (*pulAB* and *pulC* to *pulO*), plus an independently transcribed gene, *pulS*. The first step of PulA secretion, across the cytoplasmic membrane (see Fig. 1.1), is dependent upon the Sec system, the N-terminal signal sequence present in PulA, and the signal peptidase activity of LspA.[17] Once delivered into the periplasm in this GEP-dependent manner, PulA is apparently translocated across the outer membrane by a transmembrane complex composed of the products of 14 genes (*pulC-O* operon plus *pulS*). Following secretion, PulA remains transiently associated with the cell surface, before being released when cells reach the stationary phase.

An inner membrane-periplasmic complex for PulA translocation

PulE, essential for PulA secretion, is a soluble protein (non-membranous), which nevertheless can be isolated in tight association with the inner membrane.[18] PulE contains an ATP-binding motif, but shares no other sequence similarities with either SecA (the GEP protein translocase), or with members

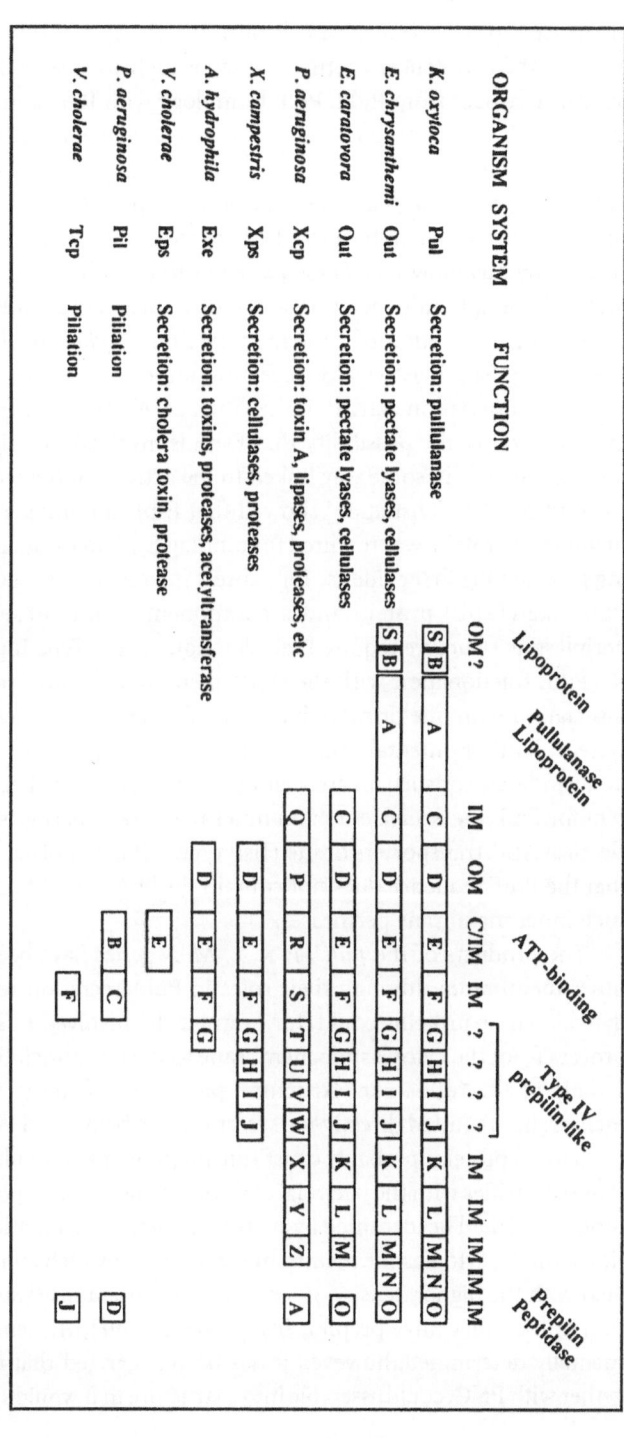

Fig. 1.2. Type II secretion systems and their homology related to the *K. oxytoca* Pul system, which is shown in the top line according to the order of genes in the *pul* cluster. Only proteins presenting a score higher than 20% overall sequence identity to its Pul partner are indicated. Proposed function and subcellular localization of the proteins are indicated above the panel: OM = Outer Membrane, IM = Inner Membrane, C = Cytoplasmic. Proteins belonging to the *P. aeruginosa* Xcp system are also called Pdd or Pil, XcpA and PilD are the same protein. PulD is equivalent to pIV of the Type VI pathway.

of the ABC protein translocase family. For example, the characteristic region between the two Walker motifs,[19] which is highly conserved in all ABC transporters, is absent from PulE. PulE homology with PilB, a protein required for the assembly of type IV pili in *Pseudomonas aeruginosa*,[20] led to the proposition that PulE could be an ATPase involved in some way in the assembly of the pullulanase secretion machinery, together with PulO (see below).[18] However, since PulE is located on the cytoplasmic face of the inner membrane, it is difficult to envisage how its ATPase activity can be utilized for the assembly of a primarily periplasmic secretion apparatus. Interestingly, a PulE homolog, involved in the secretion of cholera toxin from *V. cholerae*, EspE, was recently purified by Sandkvist et al,[21] revealing under in vitro conditions a kinase, autophosphorylation activity rather than an ATPase. As suggested by the authors this raises the possibility that EspE is involved in a signal transduction process which is in some way linked to the activity of the secretory apparatus. In addition, Sandkvist et al[21] showed that EspL, a homolog of the PulL inner membrane protein, was required for binding EspE to the membrane. This finding provides the first evidence for a direct interaction between the cytoplasmic PulE-like (EspE) protein and a component of the presumed membrane-periplasmic structure required for secretion via the Type II pathway.

PulE fractionation with the inner membrane is most probably due to its association with one or more inner membrane proteins belonging to the Pul system. PulF, an integral inner membrane protein, has been proposed as the best candidate for such an association. It was suggested that PulF could represent for PulE the equivalent of an inner membrane permease, a partner of the classical ABC transporters of the HisP type.[22] It is important to note, however, that the PulF sequence does not contain the "EAA" motif which characterizes such inner membrane permeases.[23]

The products of the *pulC, F, K, L, M, N* genes have been localized in the inner membrane, although their roles in PulA secretion remain to be elucidated.[15] These proteins could, for example, be involved in an energy transfer process from the cytoplasmic membrane to the outer membrane, or in the assembly of the "pilus-like" structure, presumed to connect inner and outer membranes required directly for secretion (see below and Fig. 1.1).

Several proteins probably constitute this pilus-like complex structure. Thus, PulG copurifies with the bacterial envelope. However, two populations of PulG molecules could be identified by sucrose gradient fractionation: one in association with the cytoplasmic membrane and another which appeared to be associated with the higher density outer membrane fractions.[24] Subcellular localization of the other three prepilin-like proteins (PulH,I,J), remains to be experimentally determined; however, it has been suggested that these proteins, together with PulG, could assemble into a structure that would span the periplasm, connecting both inner and outer membranes to facilitate the final secretion of pullulanase to the medium (see Fig. 1.7). It has been proposed that this assembly process could involve PulE, PulF and PulO and therefore require the ATPase action of PulE.[17]

PulA transport across the outer membrane

Both PulD and PulS have been identified as outer membrane proteins. PulD is tightly associated with the outer membrane while PulS is a lipoprotein, associated with the outer membrane.[25] PulD, as its homolog OutD in *Erwinia*, shows extensive homology with the pIV outer membrane protein of the filamentous phage f1.[26] OutD and pIV also show homology with YscC, involved in the secretion of virulence factors by a Sec-independent mechanism in *Yersinia pestis* (see below). This again illustrates an important sub-theme of this review, the fascinating cross-connections between different secretion pathways, as key "Lego" units are commandeered to fulfill a specific role in otherwise differing pathways.

In relation to the potential role of PulD in the exit of PulA to the medium, crosslinking experiments, followed by immunoprecipitation, showed that the PulD homolog, pIV protein, forms homomultimers composed of 10 to 12 molecules, which insert into the outer membrane of bacteria infected by phage f1.[26] This pore-like structure presumably could then be used by the new-born phage particles to leave the bacterial cell, without disturbing the integrity of the cell envelope. Interestingly, OutD and pIV are able to associate as a heteromultimers when both proteins are coexpressed in the same cell.[26] These results seem to imply that PulD homologs are able to form multimers which insert into the outer membrane as a pore to allow the secretion of specific extracellular proteins. Indeed, recent studies indicate that PulD can form multimeric, SDS-resistant complexes, fractionating with the outer membrane.[27]

Possible mechanism for Type II secretion

Both PulO, and its *Erwinia* homolog, OutO, display extensive homology with PilD/XcpA, the peptidase specific for type IV prepilins of *P. aeruginosa* (see Figs. 1.2 and 1.3). Moreover, PilD substrates contain a specific cleavage site which remarkably is also present in PulG, H, I and J, which are in fact now known to be processed and N-methylated by PulO.[15,24,28-30] Analogy with the Pil system has led to the hypothesis that an intracellular "pilus-like" structure could be involved in PulA secretion.[29] Such a structure, spanning the periplasm, would serve as a platform or scaffold for the assembly of other components, for example in the outer membrane, required for secretion, and/or to target or guide the proteins to be secreted to the outer membrane exit.[24] On the other hand the precise role of the numerous ancillary proteins required for pullulanase secretion

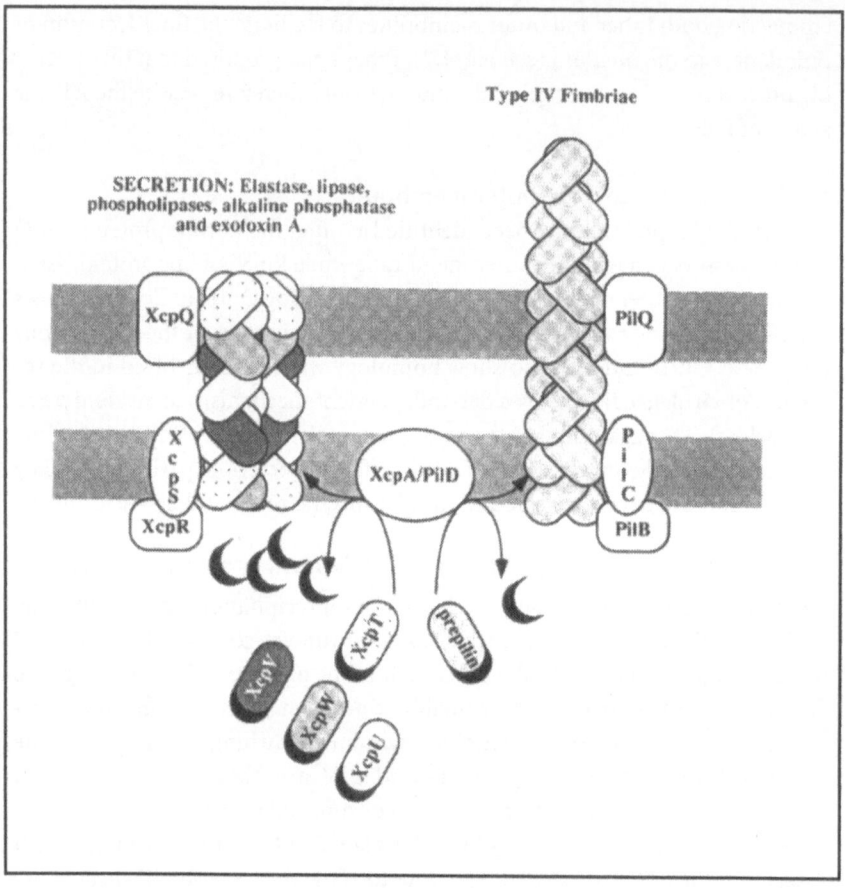

Fig. 1.3. Schematic representation of the interrelationship between the Xcp/Pil systems in *P.aeruginoa*, with proteins implicated in Type II secretion pathway (Xcp) and in pilus assembly (Pil). Reproduced, with modifications, from Hobbs and Mattick.[16]

remains elusive, in particular the PulG, H, I, J proteins and the 6 or 7 other Pul proteins which appear to associate with the inner membrane in some way (see Fig. 1.1). In particular, the need for such an "apparatus" is puzzling, since PulA is initially exported to the periplasm by the GEP. One may envisage two possibilities, not mutually exclusive, to explain the requirements for this large number of ancillary proteins. These proteins may either form a transenvelope complex required for folding (see below) and precise targeting of PulA to the PulD exit, or such a transenvelope structure, joining together the inner and outer membrane, may provide the means to tap the proton motive force of the inner membrane to provide "gating energy" to catalyze the activity of the PulD secretion complex in the outer membrane.

Finally, similarities between both Pil and the Pul/Out systems (Figs. 1.2, 1.3) are further supported by the finding that mutations in the *P. aeruginosa* *xcpA* gene (PilD) are responsible for defects not only in pilus biogenesis but also in the secretion of proteins via the Type II pathway in that organism.[24,31]

Recently, a possible function has been assigned to another component of the PulA secretory machine, the PulS lipoprotein. It has been shown that the presence of PulS, which is associated with the outer membrane of bacterial cells[25] is absolutely necessary for the stable assembly of PulD in the outer membrane. Furthermore, in the absence of PulS, the majority of the PulD protein, expressed from a IPTG inducible promoter, appeared in a degraded form. This degradation could be overcome by overexpressing PulS under the same conditions. These results indicate that not only PulD assembly but also its stabilization is dependent on the presence of PulS.[27] It is interesting to note that similar results were obtained in the case of the transmembrane complex responsible for the secretion of hemolysin A (HlyA) from *E. coli* cells (secretion Type I, see below). In this secretion system, the assembly of the inner membrane protein HlyD is dependent on the presence of TolC in the outer membrane and HlyB in the inner membrane, while HlyD itself is absolutely necessary for HlyB localization in the inner membrane.[32]

Targeting signals for Type II secretion

Conflicting results have been obtained concerning possible targeting signals for secretion across the outer membrane via the Type II pathway. Thus, Py et al[33] analyzing the secretion of the EGZ cellulase from *E. chrysanthemi* proposed that the targeting signal may be linked to the 3D structure of the protein, rather than through a specific, linear secretion signal. As described below, this would be consistent with the evidence that proteins of the Type II pathway are indeed folded *before* transport across the outer membrane. However, in some contrast with the conclusions of Py et al,[33] it has been recently reported that random insertions inside the *pulA* gene shown to abolish pullulanase catalytic activity did not affect secretion, suggesting that alterations in the structure of the enzyme do not necessarily block its targeting to the export machinery.[34]

Moreover, a discrete domain within another protein secreted by a Type II pathway, exotoxin A, has recently been identified which appears to contain all the information apparently necessary for secretion of this toxin, as well as of a passenger protein (β-lactamase), from the periplasm of *P. aeruginosa*. The extracellular targeting signal in exotoxin A is localized between residues 60 and 120 from the N-terminus. This sequence has been shown by crystallography to be rich in anti-parallel β-sheets, and in the three-dimensional view is located in the surface of the toxin, easily accessible to the components of the secretory machinery.[35] In the case of PulA, a more recent study has revealed a more complex situation. Two regions of 78 and 80 residues, one close to the N-terminus, the other much closer to the C-terminus, were clearly shown to be essential for targeting to the medium.[35a]

Proteins of the Type II pathway may fold in the periplasm prior to secretion

Evidence of the translocation of apparently fully folded proteins through the outer membrane has previously been described in the case of the multimeric protein, cholera toxin, secreted by *Vibrio cholerae*.[36] Cholera toxin is composed of five identical B subunits of 11.6 kDa and one A subunit of 28 kDa, which are exported to the periplasm via the GEP. Once processed and released in the periplasm, the mature proteins fold and assemble into an AB_5 complex.[37,38] Only when fully assembled into its final conformation is the cholera toxin apparently translocated across the outer membrane of *V. cholera.*.[36,39] The pathway used for the secretion of this protein needs further characterization, but it clearly follows a Type II secretion route.[21]

Interestingly, it has been shown that disulfide bond formation in the periplasm, involving DsbA, is an essential prerequisite for the final step in the secretion of pullulanase from *K. oxytoca*, the endoglucanase from *E. chrysanthemi* and pro-aerolysin from *Vibrio spp* through the outer membrane to the medium.[27,40,41] These results indicate that both pullulanase, EGZ cellulase and pro-aerolysin are folded in the periplasm before they are finally secreted across the outer membrane. So far there have been no reports of other periplasmic factors involved in protein folding of the Type II allocrites, but these might be anticipated to involve, for example, the bacterial PPIases[42] or the Skp protein described by Chen and Henning.[43]

SECRETION PATHWAY IV: THE "AUTOTRANSPORTERS"

In remarkable contrast to the extraordinarily complex Type II secretion pathway, proteins translocated via a type IV secretion pathway do not apparently require any auxiliary factors to complete their translocation across the outer membrane. This group of proteins includes a *Neisseria gonorrhoeae* and *Haemophilus influenzae* immunoglobulin specific protease, IgA,[44-46] other specific proteases secreted by different species of *Neisseria*[47] as well as the serine protease of *S. marcescens*, SSP.[48] In addition, the VirG protein of *Shigella*, which promotes actin polymerization[49] and the vacuolating cytotoxin of *Helicobacter pylori*, VacA,[50] are also secreted via a Type IV mechanism. These proteins surprisingly rely for their secretion on functions present within the allocrite itself (Figs. 1.1 and 1.4).

The *N. gonorrhoeae* IgA protease secretion system is the best studied example of the Type IV secretion pathway (for a review see ref. 45). This protein

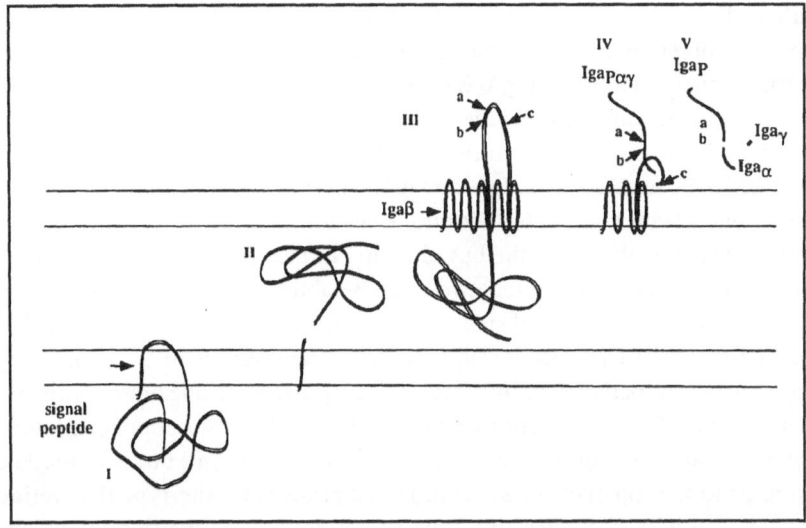

Fig. 1.4. Schematic model proposed for IgA protease secretion. Steps I+II reflect signal peptide directed export of the polyprotein (IgA$_{\pi\alpha\beta\gamma}$) into the periplasm (Sec-dependent steps); (III) insertion of the β-domain into the outer membrane, forming the pore through which the IgA$_{\pi\alpha\gamma}$ domains are translocated to the external medium; (IV) autoproteolytic uncoupling of IgA$_{\pi\alpha\gamma}$ from the membrane associated IgA$_\beta$ domain, upon cleavage at site c; (V) further sequential autoproteolysis at sites a and b, producing the mature IgA protease, the α-protein and the γ-peptide. Reproduced with modifications from Klauser et al.[45]

is synthesized as a 169 kDa precursor, the polyprotein IgA, containing five distinct functional domains: (i) an N-terminal domain corresponding to the signal peptide which allows transport through the cytoplasmic membrane; (ii) a core region of 106 kDa containing the protease activity of the mature protein (IgA$_p$); (iii) a small 12 kDa region, the α-domain (IgA$_\alpha$), a short (30 aa) γ-domain (IgA$_\gamma$) and (v) a C-terminal β-domain (IgA$_\beta$).

As shown in Figure 1.4, the first step in IgA$_p$ secretion is dependent upon an N-terminal secretion signal, recognized by the Sec-system and cleaved as the polyprotein is translocated to the periplasm.[48] Once in the periplasm, the C-terminal IgA$_\beta$ domain integrates into the outer membrane forming a pore which apparently then allows translocation of the domains IgA$_p$, IgA$_\gamma$ and IgA$_\alpha$ to the extracellular medium. Moreover, the IgA$_\beta$ domain appears to carry all the transport specificity since Klauser et al[47] have demonstrated that a heterologous protein (*Vibrio cholerae* toxin B subunit) fused as a passenger of the N-terminal of IgA$_\beta$ is efficiently transported across the outer membrane onto the cell surface.

Integration of the C-terminal IgA$_\beta$ domain into the outer membrane is presumed to be a rapid process since no accumulation of periplasmic intermediates can be detected under conditions of normal secretion.[45] Interestingly, Klauser and co-workers have argued, based on secondary structure predictions, that the key structural feature of IgA_β is likely to be a β-barrel, which is characteristic of most integral outer membrane proteins.[47]

After being translocated to the periplasm, the N-terminal domain of IgA destined to be secreted (IgA_p, IgA_γ and IgA_α) must presumably be maintained in an unfolded state, compatible with translocation through the channel formed in the outer membrane by the IgA_β domain. In fact, premature folding in the periplasm of a fusion protein between the B-subunit of cholera toxin and IgA_β ($CtxB$-IgA_β) blocks the final step of secretion through the outer membrane. The efficiency of translocation is restored in the presence of the denaturing agent β-mercaptoethanol or by using CtxB derivatives, lacking one or both cysteines involved in the formation of disulfide bonds.[47] Translocation of unfolded intermediates is of course characteristic of the Sec-system[51] but is in marked contrast to the apparent transport of folded proteins via the Type II secretion pathway. In turn, therefore, this Type IV pathway raises questions about the mechanism of refolding of secreted IgA protease, in the medium following secretion. This problem will be referred to in later discussion of the Type I pathway.

Once translocated through the IgA_β pore, $IgA_{p\gamma\alpha}$ adopts its active conformation and is released by autoproteolysis from the IgA_β domain, which remains attached to the outer membrane. Release of the protease into the medium is the result of initial cleavage at site *c* (Fig. 1.4) Extracellular $IgA_{p\gamma\alpha}$ is then further processed by sequential autoproteolysis at sites *b* and *a*, to produce the mature IgA protease, the α-protein and the small γ-peptide corresponding to the intervening 30 residues between cleavage sites *a* and *b*.[47]

The precise function of the alpha protein (IgA_α) is still unknown. However, this question has been addressed in an homologous system, responsible for the secretion of the serine protease (SSP) from *S. marcescens*. In this system, a small peptide of 71 residues, called the "junction region," is also localized between the central active domain of the protease and the C-terminal domain responsible for its final secretion. Trans-complementation experiments have shown that this junction region, which remains attached to the outer membrane, surprisingly is essential for the activity of the secreted SSP, suggesting that this region plays a role in the final folding of the mature protein. This result indicates the existence of a third activity for these polyproteins, as "intra-molecular chaperones," capable of ensuring an autocatalytic, correct folding of the final secreted protein.[52] However, the actual mechanism by which this junction region acts as an intramolecular chaperone for the mature protein remains to be established.

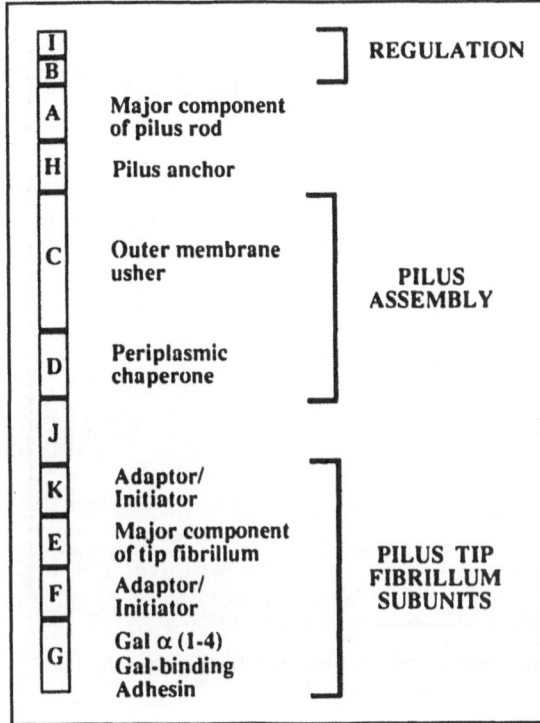

Fig. 1.5. The pap cluster, responsible for the production and secretion, via a Type V pathway, of the *E. coli* P-pilus subunits.

SECRETION PATHWAY V: THE *E. COLI* PAP SYSTEM

The *E. coli* Pap system is responsible for the translocation and assembly of P pili (fimbriae) onto the external surface of the bacterial outer membrane (for a detailed review see reference 53). Several proteins constitute the final cell surface pili (Fig. 1.5) which requires a specific assembly apparatus (see Fig. 1.6). It is important to note that P pili are distinct from *Pseudomonas* pili which are secreted via a different mechanism, a variation of the Type II pullulanase secretion pathway. P pili are composed of two distinct structures, the rod and the tip fibrillum, that extends from each pilus rod. The rod is formed by repeating subunits of PapA arranged in a right handed-helical cylinder. The fibrillae are mostly composed of subunits of PapE arranged in an open helical conformation, with the PapG adhesin, responsible for the specific recognition of the Pap pilus receptor, at the distal end of each fibrillin tip. PapF and PapK are necessary to ensure that the assembly of PapG adhesin precedes that of the pilus rod (PapA), and that all pilus subunits are firmly connected.[54] A single copy of another pilus subunit, PapH, is found on the base of the growing pilus, and it is

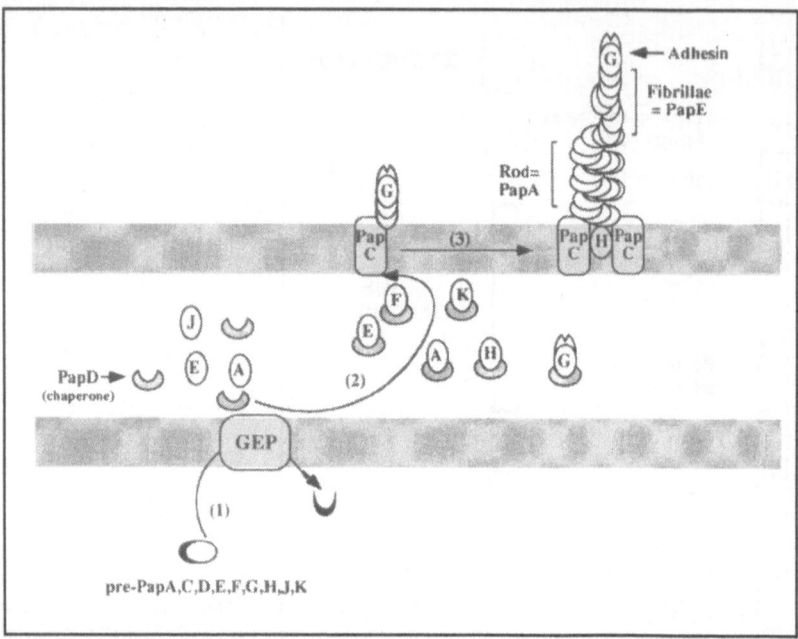

Fig. 1.6. Model for Pap pili subunits secretion and assembly through the PapC "platform". Modified from Hultgren et al.[53]

assumed that its role is to prevent further incorporation of other subunits, thus controlling pilus length, and to anchor the pilus to the assembly platform. All these Pap proteins are secreted to the outside of the outer membrane and two other proteins, PapC and D, respectively an outer membrane usher and a periplasmic chaperone, are responsible for the correct secretion and then assembly of the six subunits into a functional pilus structure.[6,53] PapJ, an 18 kDa periplasmic protein showing homology with nucleotide-binding proteins, is responsible for maintaining the integrity of the pilus structure. Mutations in *papJ* result in large amounts of pilus antigen being released from the *E. coli* cells, apparently resulting from internal breaks in the pilus. It has been suggested that PapJ could function as a chaperone, participating directly or indirectly in the correct assembly of PapA subunits.[55]

As indicated in Figure 1.6, all the pilin subunits are synthesized as precursors possessing a typical N-terminal signal sequence, and are exported through the cytoplasmic membrane via the GEP. After cleavage of the signal peptide by LepB (the signal peptidase), the pilus subunits are released into the periplasm where they form a complex with the specific chaperone PapD. Although it is not clear when each subunit folds into its final conformation, it has been shown that pilin subunits, associated to PapD, have disulfide bonds[56] and are recognized by many, but not all, monoclonal antibodies that bind to the native pilus.[57] Furthermore, it has been shown in vitro that PapG adhesin, purified from the periplasmic fraction complexed with PapD, retains its ability to bind the receptor-ligand, and is protease resistant.[58] These results indicate that pilins are close to their final conformation, if not fully folded, before secretion across the outer membrane.

From the periplasm, the pilin subunits are escorted, in chaperone-complexed form, to the pilus assembly platform, composed, it is supposed, of many molecules of the PapC outer membrane protein, which represents the translocator responsible for secreting the pilins through the outer membrane. To form the organized pilus structure, the different subunits, released from PapD, are secreted and assembled sequentially by the PapC assembly platform, in a rapid process (less than 5 min) that seems to be independent of cellular energy and is thermodynamically driven.[59]

As indicated above the assembly of the pili of the *E.coli* Pap system is different from that of Type IV pilins in *P. aeruginosa, N. gonorrhoea* and several other Gram-negative bacteria.[16] Furthermore, although Pap pilins might appear to be similar to Type IV pilins, their N-terminal signal sequences are not processed at the same site or by the same peptidase.[15,16,60]

Secretion pathway Va: variations on a theme

At least four relatively simple two step secretion systems can also be classified, for the moment, as Type V secretion pathways. As illustrated in Figure 1.1, these apparently involve a single outer membrane accessory protein and are responsible for the secretion of the calcium-independent (non-RTX) hemolysins of *Serratia marcescens*, ShlA,[61] and *Proteus mirabilis*, HmpA,[62] the *Bordetella pertussis* filamentous hemagglutinin, FhaB[63] and possibly the *E. coli* extracellular heat-stable enterotoxin, ST_B.[64] Nevertheless, since the accessory proteins in these systems are not homologous with those required for pilin secretion, we propose to distinguish them as Type Va.

Final secretion of ShlA, HmpA and FhaB toxins is accomplished, as in the Type V Pap system by a single outer membrane protein, respectively ShlB, HmpB and FhaC, encoded by genes adjacent to those encoding the toxins themselves. In the case of ST_B, it has been suggested that the outer membrane protein responsible for its final secretion could be TolC.[64] These outer membrane transport proteins are not, however, related to PapD and, at least in the case of ShlB, appear to have an additional role in the secretion process—to activate the secreted protein (see below).

The ShlA protein itself is synthesized as an inactive precursor which is translocated across the cytoplasmic membrane by the GEP. The ShlA periplasmic form is inactive and activation seems to occur concomitantly with secretion through the outer membrane via ShlB. Periplasmic ShlA can also be activated in vitro upon incubation with a cell lysate containing ShlB, which indicates that ShlB is not only responsible for ShlA secretion but also for its activation. Such an activation might occur by a covalent modification of the inactive toxin during or after secretion, since once activated, ShlA remains hemolytic even after removal of ShlB.[65] An exciting possibility is that ShlB in some way acts as a chaperone, determining the final active conformation of the toxin during secretion. Finally, the secretion signal responsible for targeting ShlA to the outer membrane translocator has been localized to the first 238 residues of the N-terminal of the toxin. Similarly, this region of the molecule appears to be essential for penetration of the erythrocyte membrane by the active toxin.[66] As in the case of the other Type V proteins, it is not clear whether ShlA, HmpA, FhaB or ST_B fold in the periplasm, during translocation, or following release to the medium. However, at least in the case of ST_B it has been shown that DsbA is required for secretion. One may anticipate that if such proteins are already folded, the outer membrane protein, as is the case of PulD, and the pIV protein involved in filamentous phage secretion would have to be present in large oligomeric structures , but this has not so far been reported.

II. ONE-STEP SECRETION PATHWAYS

Three secretion pathways, Types I, III and VI, are used to translocate proteins directly from the cytoplasm to the external medium, bypassing the periplasm. Secretion via a Type I system, for example, hemolysin secretion (see below), is dependent upon transmembrane complexes composed of three proteins specific to the allocrite. The total number of proteins implicated in Type III secretion systems is not yet determined, although it is clear that this secretion pathway combines characteristics of systems I and II. Secretion pathway VI is mainly used to "secrete" encapsulated particles of filamentous phages, and will only be considered briefly in this review.

SECRETION PATHWAY III: THE YERSINIA YSC SYSTEM

We shall use the Ysc system to illustrate first the principles of the widespread Type III secretion pathway. The Ysc system has been identified as responsible for the secretion of a set of proteins involved in pathogenicity in *Y. pestis, enterocolitica* and *pseudotuberculosis*.[67] These bacteria attempt to escape the defense system of the host cell by secreting a V-antigen and at least 11 other proteins called Yops (*Yersinia Outer membrane Proteins*). The Yop proteins were initially detected in association with the outer membrane, but it was later demonstrated that they can actually be secreted into the extracellular medium.[68] A simple model illustrating the participation of some of the proteins involved in

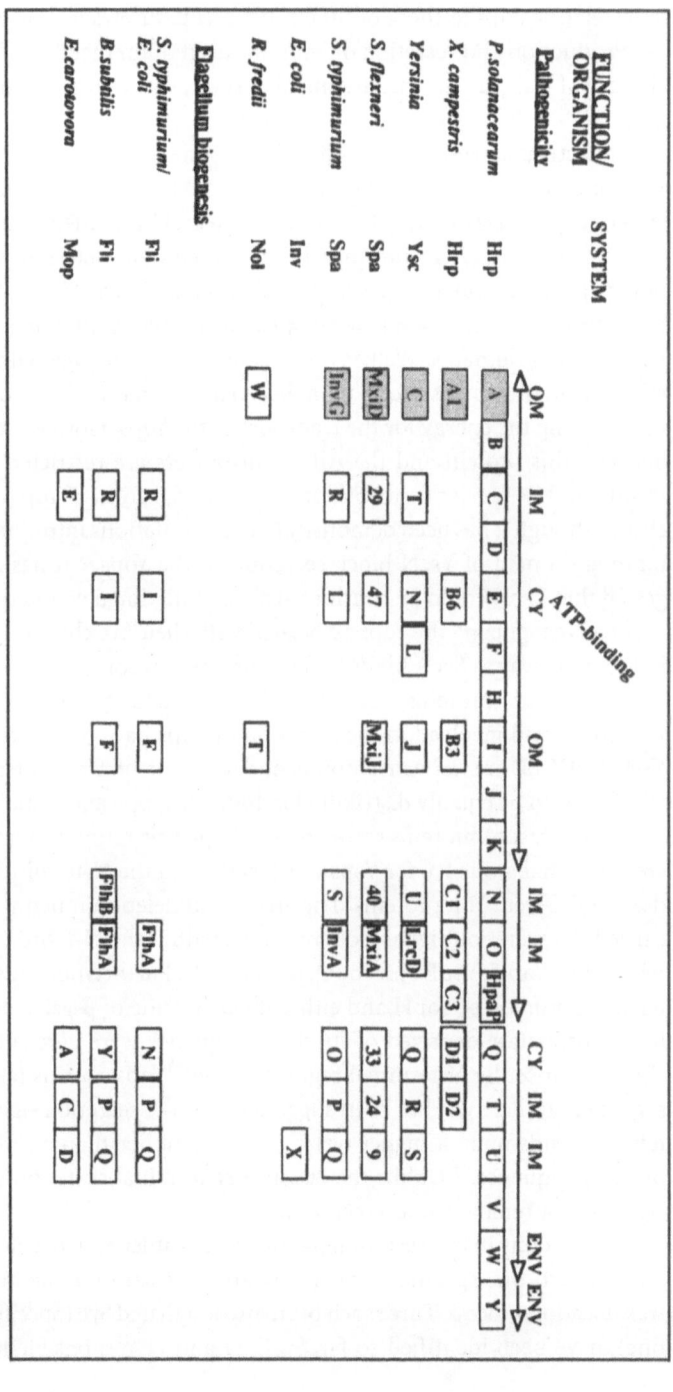

Fig. 1.7. Major homologies among determinants involved in Type III secretion pathways in different organisms. *P. solanacearum* determinant, arbitrarily selected as a prototype, has its transcription units, relative position of genes and subcellular localization of proteins indicated on top of the panel: OM = Outer Membrane; IM = Inner Membrane, CY = Cytoplasmic; Env = Envelope. ▨ = proteins sharing homology also with PulD superfamily (secretion Type II).

the secretion of Yops is indicated in Figure 1.1. The corresponding genes and their relationships to those of others Type III pathways is shown in Figure 1.7.

Production and secretion of Yops, induced upon incubation at 37°C in the absence of Ca^{2+}, are under the control of several genes present in a 70 kbp plasmid, pYV.[68] The *yop* genes are arranged in three loci, constituting a thermoactivated regulon controlled by the gene *virF*.[69] The *virC* locus contains 13 genes, organized in an operon, *yscA* to *yscM*.[70,71] The *virA* locus contains at least one gene, *lcrD*, coding for a 78 kDa integral inner membrane protein predicted to contain eight transmembrane domains, as determined by PhoA topology analysis, and a large cytoplasmic domain.[72] *yscN* is located in the *virB* locus and codes for a 47.8 kDa cytoplasmic protein containing two consensus ATP-binding domains, Walker boxes A and B. By analogy with the secretion mechanism of Type I systems it has been suggested that YscN could play a role in providing the energy for the transport of the Yops. However, the similarities between this protein and the ABC transporters are restricted to the Walker motifs, and its precise contribution to the energization of Yop transport is unclear, although it has been demonstrated that mutations introduced in the ATP-binding domain of YscN block secretion of the Yops.[73] It has also been suggested that YscN could be considered as the functional counterpart of SecA in *Yersinia*, recognizing the Yops associated with their Syc chaperones (see below) in the same way as SecA binds to SecB-preprotein complexes.[73]

Three other proteins, YscC, D and G, encoded by genes belonging to the *virC* locus and involved in the secretion of Yops have been recently localized. YscC and YscD are respectively outer and inner membrane proteins, while YscG was found to be equally distributed in both envelope and soluble fractions.[74]

In contrast to proteins secreted via a Type I secretion system, a specific secretion signal sequence for Yops was localized to the N-terminal 48 to 98 residues of different Yop proteins. The analysis of deletion mutants has in fact localized the secretion signal sequence to be within the N-terminal, 48, 98 or 76 residues of YopH, YopE and YopQ, respectively. Furthermore, fusions of the N-terminal domain of YopH and either the α-peptide of β-galactosidase or alkaline phosphatase deprived of its signal sequence were efficiently secreted by *Yersinia*. Since the N-terminal regions of these Yop proteins fail to reveal any sequence similarity, it has been suggested that the secretion signal of Yop proteins depends upon a higher order structure rather than a simple sequential primary sequence.[75] Unlike the classical N-terminal secretion signal, the Yop signal is not removed upon secretion.[75]

Remarkably in this secretion pathway, ATP-independent cytoplasmic chaperones, each one specific to each Yop, are necessary for the initiation of the translocation process. Three such proteins, designated Syc (*S*pecific *Y*op *C*haperone), have been identified so far, SycE, H and D, responsible for the specific secretion of YopE, H and D, respectively.[76,77] Similarly, dedicated chaperones for *Shigella flexneri* Ipa invasins were demonstrated.[78] It has been suggested that the Syc proteins (or IpgC in *S. flexneri*) are necessary to avoid aggregation of different Yop or Ipa proteins in the cytoplasm.[77,78] This may be important since different Yop or Ipa proteins may be required to form specific complexes following secretion. Alternatively, Syc proteins may be essential to target the Yops from the cytoplasm to the membrane localized secretory machinery. However, the precise role of these chaperones in the secretion process (reviewed in detail recently by Wattiau et al[79]) remains to be established.

The first step in the translocation process of Yops from the bacterial cells is probably dependent on the YscN ATPase and on the inner membrane proteins LrcD, YscD and YscG (see Fig. 1.1). YscR and YscS are probably inner membrane proteins and remarkably, together with LrcD, with which they presumably interact, these proteins share homology with proteins involved in flagellar biogenesis. In fact as indicated in Figure 1.7, at least 7 Ysc proteins share homologies with proteins involved in flagellar biogenesis. Similarly, a recent more detailed study (see also Fig. 1.7) has identified the same set of conserved proteins in the Type III pathway in the plant pathogen, *Pseudomonas solanacearum*.[80]

The obvious homology of some elements of the Type III pathway with components of the transenvelope structure constituting the complex flagellar motor system has given rise to the idea illustrated in Figure 1.1 that secretion of Yops also involves a transenvelope structure. Presumably in this case, as in Type I secretion, it is accessible from the cytoplasm.

The only identified outer membrane component of the translocation machinery, YscC, is a homolog of PulD, OutD and pIV (secretion pathways, Types II and VI), and this protein is proposed to be involved in the last steps of translocation, possibly constituting a pore in the outer membrane for final export to the medium.[74] Again this is a remarkable example of otherwise different pathways using a conserved protein at a common stage of the secretion process. It may be relevant to recall therefore that at least OutD and pIV probably function as large oligomeric channels for secretion of proteins already folded. The requirement for chaperones, possibly as anti-aggregation factors for Yops, may indicate that these proteins are also folded before translocation, in which case large oligomeric forms of YscC might be anticipated.

Type III secretion, a ubiquitous pathogenicity element

Similar Type III systems (see Fig. 1.7), responsible for secretion of virulence factors, have been identified in different pathogenic bacteria, such as *Shigella flexneri* , designated Mxi-Spa (responsible for the secretion of Ipas or invasins),[81,82] *Salmonella typhimurium* (Inv),[83] enteropathogenic *Escherichia coli* (EPEC - Sep system),[84] *Xanthomonas campestris* and *Pseudomonas solanacearum* and *seringae* (Hrp).[85,86] Details of the *Salmonella* and *Shigella* systems have been reviewed in a recent publication, with emphasis on the fact that secretion in these cases may well be linked to close contact between bacterial and host cells, allowing efficient translocation of certain proteins from bacteria "directly" into the cytoplasm of the host.[87-89]

At least 10 bacterial species, ranging from animal to plant pathogens, have in fact now been identified, expressing this highly conserved protein secretion mechanism, coupled to uptake into target cells. As discussed in a recent review, fascinatingly the proteins of the Type III secretion system and the toxic proteins, which they discharge into target cells, appear to have been acquired by bacteria more than 4 million years ago from an "unidentified source."[90]

SECRETION PATHWAY VI: BACTERIOPHAGE PARTICLE EXTRUSION SYSTEMS

This is the secretion pathway used by large encapsulated phage particles, such as M13 or F1, to reach the extracellular medium without disrupting the integrity of the bacterial host cell (reviewed in reference 91). The assembly and extrusion of F1 phage particles are coupled events initiated in the bacterial inner membrane and involving the phage DNA, and viral coat proteins. In addition, the mechanism requires at least one host protein and two phage-encoded proteins, pI and pIV, which are implicated in the formation of a trans-envelope structure that allows direct "secretion" of phage particles to the medium;[26,91] pI is an inner membrane protein with a large cytoplasmic and a small periplasmic domain, separated by a single transmembrane region.[92-94]

pIV has been localized to the outer membrane, but can also be detected in association with the inner membrane.[95,96] Coimmunoprecipitation, crosslinking and sedimentation experiments indicate that the structure of the outer membrane pore responsible for the final secretion of f1 phages may be composed of 10 to 12 monomers of pIV protein. It has been proposed that pI and pIV form a transenvelope structure that allows direct secretion of f1 phage particles to the medium. Since no interaction between pI and pIV could be demonstrated so far, it has been suggested that such a transenvelope structure is formed only transiently, possibly coupled to the initiation of the phage assembly complex, which involves the cytoplasmic domain of pI.[26] Interestingly, as discussed above, pIV is one of the PulD homologs, which are themselves involved in the last step of the GEP-dependent Type II pathway (section 1.1).

IV. HEMOLYSIN SECRETION FROM *E. COLI*

Secretion of a "hemolysin" by certain pathogenic strains of *E. coli* was first recognized more than 90 years ago by Kayser et al.[97] However, we had to wait until the early 1980s before the molecular mechanism of hemolysin (HlyA) secretion from *E. coli* began to emerge. Throughout this period and until recently it was assumed that HlyA was the only leaderless protein specifically secreted by *E. coli*. Kenny and Finnlay[98] have now shown that at least 5 proteins are secreted by EPEC (enteropathogenic *E. coli*) strains in the presence of epithelial cells. The mechanism of secretion of these proteins is however not established.

In a book concerned with protein secretion by unusual routes, hemolysin (HlyA), the protein probably studied in the most detail, clearly deserves an important place. Happily for the groups who study it, the Type I secretion pathway followed by HlyA is relatively simple compared to most other pathways. Many aspects of hemolysin secretion, in particular concerning the genetic organization of Hly-determinants, regulation of expression of *hly* genes and activation and mode of action of hemolysin itself have been reviewed previously and will not be covered here. Rather, we shall attempt to concentrate upon the last five years where more specific detail of the HlyA secretion signal, the nature of the translocator and the mechanism of secretion have appeared. At least 12 additional examples of the Type I secretion pathway (see Fig. 1.8) have now been described and, where appropriate, these other systems will be introduced, but most emphasis will be placed on the *E. coli hly* system.

BASIC ESSENTIALS OF THE *E. COLI* HEMOLYSIN SECRETION PATHWAY

Secretion of hemolysin is only observed in some uropathogenic *E. coli* strains with specific *hly* genes encoded chromosomally in human pathogens or on plasmids in strains found in many animal species. As shown in Figure 1.8, *hlyA* determinants invariably contain the genes, *hlyC*, encoding a 20 kDa cytoplasmic protein, the *hlyA* structural gene encoding a 110 kDa protein, and *hlyB*, *hlyD* encoding two inner membrane proteins of 79 kDa and 53 kDa, respectively. HlyC, together with the acyl carrier protein, Acp, promotes specific fatty acid modifications of HlyA, essential for toxin activity.[99] HlyC plays no part in secretion,[100] while HlyB and HlyD are absolutely essential for secretion, probably forming a transenvelope translocation complex with TolC, which is also essential for secretion.[101] Secretion of HlyA does not depend upon the SecAY protein export pathway. In contrast, a novel C-terminal secretion signal is essential for an apparently single step transport mechanism, which takes HlyA directly to the medium without a periplasmic intermediate.[2,102]

Secretion of hemolysin normally peaks in the late exponential phase of growth although the basis for this regulation is not clear. HlyA itself is a pore forming toxin, whose activity is dependent upon Ca^{2+}.[103] In fact, HlyA is a member of the RTX family of proteins characterized by the presence of tandemly repeated, glycine-rich nonapeptides in the C-terminal domain, as shown in Figure 1.9. In the case of HlyA there are 13-14 such repeats located downstream of the membrane binding region and approximately 150 residues proximal to the secretion signal.

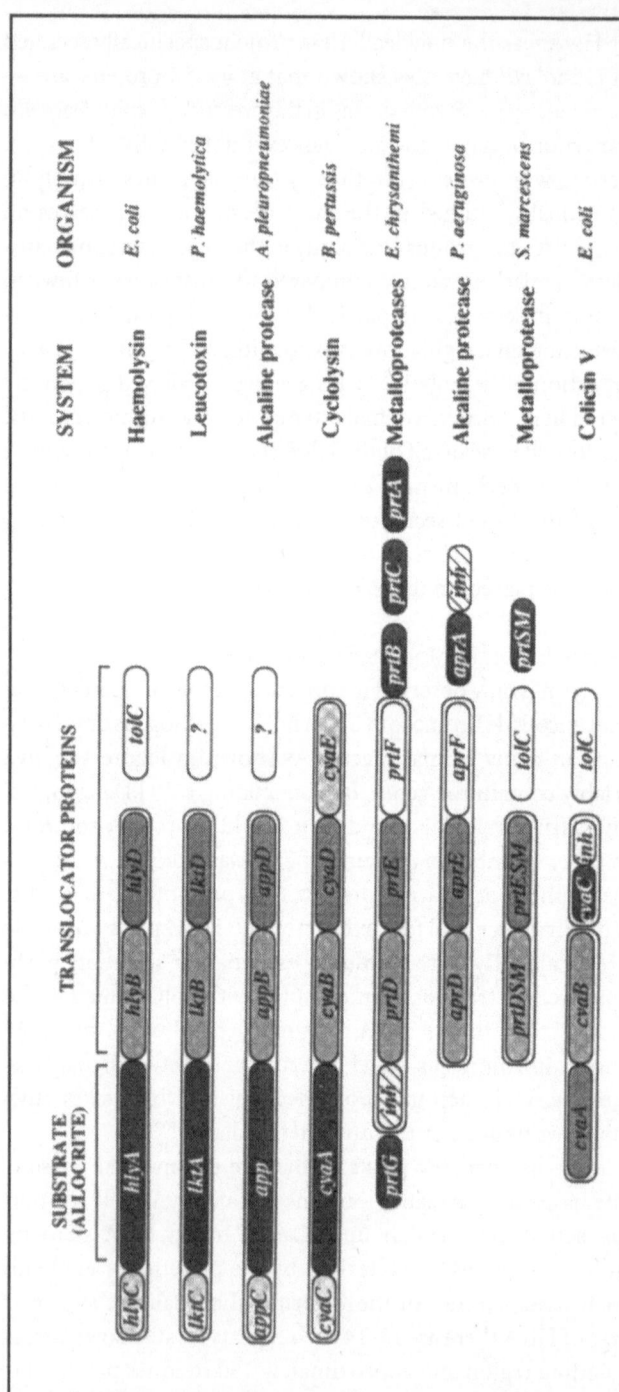

Fig. 1.8. Schematic representation of the organization of different determinants homologous to hly20001. ▮ =Allocrite; ▮ = ABC component; ◯ = HlyD (MFP) component; ◯= outer membrane component; ▮ = toxin activator; ▨ = toxin inhibitor.

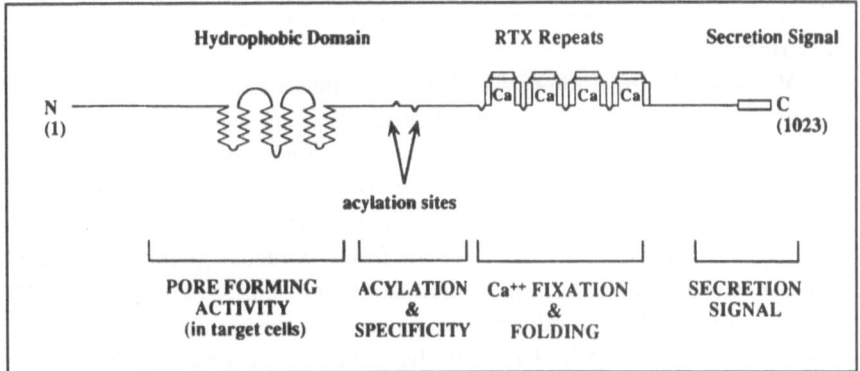

Fig. 1.9. Functional domains of the hemolytic cytotoxin HlyA. Reproduced with modifications from Ludwig and Goebel.[97]

In succeeding sections we shall describe in more detail the nature of the HlyA secretion signal and the structure and possible functions of the HlyB, D and TolC components of the translocator. Finally, we shall discuss the possible mechanism of HlyA secretion.

ANALYSIS OF THE HLYA C-TERMINAL SECRETION SIGNAL

Minimal size

Several lines of evidence have demonstrated unequivocally the presence of a specific secretion signal at the C-terminal of HlyA. Thus, deletion of more than 90% of the protein from the N-terminus of HlyA still allows the remaining C-terminal fragment to be secreted, while C-terminal deletions of only a few amino acids can greatly reduce secretion. The smallest reported C-terminal peptide of HlyA which can still be translocated in the presence of HlyB and HlyD is 62 amino acids.[104] With the RTX-protein PrtB from *E. chrysanthemi*, efficient secretion of a peptide containing only the 29 C-terminal amino acid residues has been reported.[105]

Strong evidence for a C-terminal secretion signal for HlyA was obtained from its ability to promote the secretion of a variety of heterologous passenger proteins, specifically in the presence of HlyB, D.[106,107] The smallest C-terminal peptide shown to promote heterologous protein secretion was also 62 residues.[104] However, secretion efficiency is highest when larger C-terminal fragments up to 213 residues are employed.[106] Similar size dependence effects on secretion of passenger proteins were obtained with the PrtB system.[108] As discussed below, we conclude from these results that the secretion of some passengers may be inhibited by occlusion of the signal by the upstream sequences, and not because the secretion signal of HlyA and other RTX-proteins extends beyond 40 to 60 C-terminal residues.

Various deletion studies have also shown that a specific secretion signal occurs at the extreme C-terminal of other RTX-proteins including leukotoxin[109] and adenyl cyclase toxin, although in the latter case evidence was provided that alternative secretion signals, perhaps normally latent, were present between approximately 100 and 300 residues from the C-terminal.[110]

Extensive deletion and fusion studies from Wandersman and co-workers have shown that the C-terminal secretion signal of a metalloprotease (PrtB) from *E. chrysanthemi* is minimally contained within the terminal 29 residues.[105] Moreover, Ghigo and Wandersman[105] have shown that the C-terminal motif DIVV is particularly important for function and that even a single residue addition to the C-terminal blocks secretion.

However, these other signal sequences have not been subjected to detailed genetic analysis. Notably, alignment of the C-terminals of these other proteins of the Type I pathway, with that of HlyA, indicate only limited conservation of primary sequences (see below and Fig. 1.10)

Detailed mutational analysis of the HlyA secretion signal

The C-terminal region of HlyA has been subjected to extensive mutational analysis with many single and multiple residue substitutions identified. These results are consistent with a specific secretion signal extending over the entire 46 C-terminal residues but perhaps no further. However, within this region, many residues can be substituted without significant loss of activity indicating that the signal is highly redundant. In contrast, at least seven residues have been identified whose substitution leads to a substantial (50-70%) loss of secretion efficiency.[111-113] Such residues appear to be relatively dispersed throughout the signal with a possible functional hot spot (EISK) at the extreme proximal end (residues -43 to -46).[113]

As shown in Figure 1.10, the main primary sequence motifs and predicted secondary structural features of the HlyA C-terminal are a Helix (Helix II) turn Helix (I), a charged region, DVKEER overlapping Helix I and an 8-amino acid C-terminal enriched for hydroxylated residues. These features are, however, only conserved in the closest relatives of HlyA. In particular, the charged region and hydroxylated tail are absent from other RTX-proteins. Single residue mutagenesis and deletion studies also failed to support a major functional role for the charged residues in the DVKEER motif.[112,114] Similarly, insertion of a helix breaker, Pro, or severe truncation of Helix I do not have major effects on secretion efficiency.[112,114] In addition, recent site directed mutagenesis failed to confirm a major functional role for Helix II.[113] In contrast, two earlier studies have emphasized the crucial importance of the relatively well conserved Phe (F989)

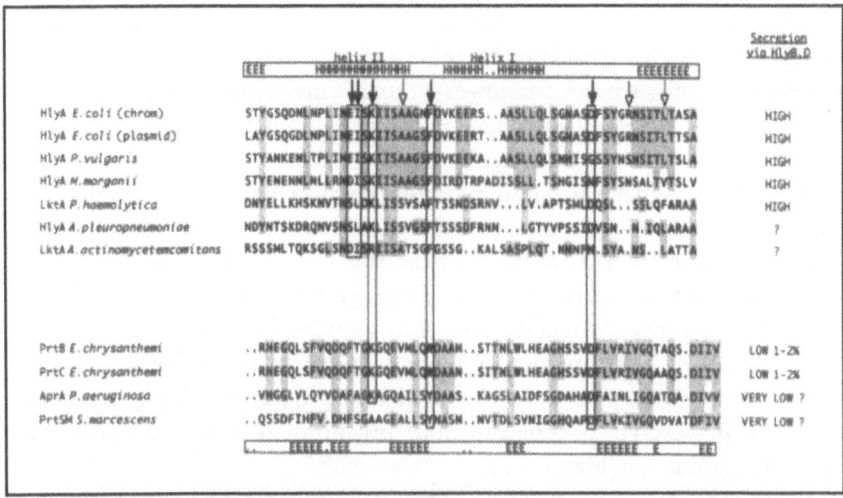

Fig. 1.10. Alignment of the C-terminal residues of RTX-proteins secreted via the Type I secretion pathway. The predicted secondary structures for the HlyA-like and PrtB-like proteins are indicated at the top and bottom respectively. Highly conserved residues are indicated in vertical boxes. Solid and open arrows indicate important functional residues identified by mutagenesis by Holland and colleagues[112,113,115] or Koronakis and colleagues,[111] respectively. At the right is indicated the level of secretion of each RTX-protein by the Hly translocator relative to secretion by the homologous translocator as indicated in the literature.

residue (-35) in the predicted turn region between Helix II and Helix I.[112,114] Chervaux and Holland[113] have now shown that many substitutions of F989 result in a substantial loss of function, with a Pro substitution giving only 5% secretion of HlyA to the medium.

From all these studies and in consideration of physical studies of the signal region of RTX toxins (see below), we conclude that several, dispersed individual residues, including F989 and a cluster close to E978, are essential for recognition and/or docking with the HlyBD translocator, perhaps largely independent of specific secondary structure.

Is the HlyA secretion signal bifunctional?

The novel HlyA secretion signal and the C-terminal secretion signal of other RTX-proteins are quite distinct from N-terminal signal sequences both in size, amino acid composition and overall size. Although it has been suggested[111] that the HlyA signal may initially interact with the lipid bilayer, no evidence has yet directly supported this, and we prefer the more economical hypothesis that the signal sequence interacts directly with the HlyBD translocator, in a specific *recognition* interaction. We have previously described a LacZ-HlyA fusion (carrying a wild-type secretion signal) which forms an abortive translocation complex with HlyBD.[106] This hybrid protein competes with coexpressed wild-type HlyA for secretion and such competition experiments, involving fusion proteins with wild-type and mutant signals, demonstrated that mutations in the region -15 to -46 were recessive, i.e., no longer competing. We proposed that this proximal portion of the signal is involved in translocator recognition.[115] In contrast, we have now identified a dominant mutation (as defined by competition experiments) at the extreme C-terminal of the HlyA signal, which also appears to reduce the rate of secretion.[116] More strikingly, this mutation also blocks the formation of active toxin and increases its sensitivity to trypsin (this lab, unpublished data). The properties of this mutant indicate that the distal region of the secretion signal may also be involved in a final release/folding step in the secretion process. The possible functional organization of the signal peptide of HlyA is illustrated in Figure 1.11.

Structural studies of the secretion signal of HlyA and other RTX-toxins

Baumann and co-workers have published X-ray crystal structures of two RTX-toxins, proteases secreted by *Pseudomonas aeruginosa* and *Serratia marcescens* respectively, which clearly show the largely β-strand structure of the C-terminal signal region.[117,118] Algorithms also predict that these regions are rich in β-strand as shown in Figure 1.10. This figure also shows that RTX-toxins can be subdivided into at least two subfamilies, HlyA-like and PrtB-like. With the exception of the extreme C-terminal, the HlyA signal is predicted as largely α-helical. In contrast to the X-ray studies, an NMR analysis of the signal region of PrtB indicated a helical structure.[119] Similarly, an NMR study of the HlyA and LktA signal peptides indicated a structure containing two α-helices.[120] However, in both studies helical structures were detected for HlyA, LktA and PrtB only in the presence of organic solvents, while the peptides were unstructured in aqueous solution. Similarly, the HlyA signal peptides were unstructured in aqueous solution when analyzed by CD-spectroscopy (our unpublished data and reference 121). The obvious discrepancy between the crystal structure of some PrtB-like secretion signals and the NMR analysis of the PrtB protein in TFE suggest strong caution in interpreting these peptide structures obtained in organic solvents. Again in our view, as supported by the bulk of genetic evidence, we prefer the idea that, in vivo, relatively unstructured or unfolded signal peptides recognize or dock with the translocator involving a few specific residues. This view does not conflict with the likelihood that *after* translocation, signal peptides fold into β-strands or helical forms, according to subfamily, upon emergence from the translocator. Indeed, this may be an essential step in the final folding/release of the protein into the medium.

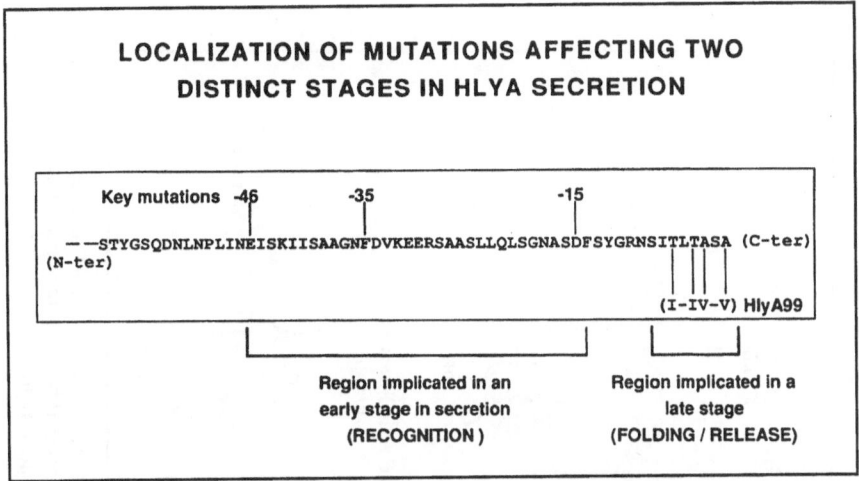

Fig. 1.11. Mutagenic analysis of the HlyA C-terminal secretion peptide has identified two possible functional roles.

COMPONENTS OF THE TRANSLOCATOR—HLYB

HlyB is the prototype ABC transporter first identified by sequencing of the gene by Welch and colleagues in 1986.[122] HlyB (79 kDa) is composed of an approximately 30 kDa cytoplasmic ATP-binding protein directly linked to an N-terminal domain forming an integral membrane protein. This domain is however not strongly hydrophobic and probably contains extensive inter-transmembrane domain (TMD) loops, making topological predictions difficult, a feature typical of the ABC transporter superfamily which has been reviewed in detail elsewhere.[23] Experimental topological analysis (see Fig. 1.13) using β-lactamase and other fusions indicates a minimum of 6 TMDs, as predicted for many members of the family with possibly two additional TMDs in the N-terminus of HlyB, which is larger than that of many typical members of the family. HlyB is absolutely essential for hemolysin secretion with the presumption that a highly conserved cytoplasmic ATPase energizes transport via the membrane domain. The distal portion of the membrane domain shares small but significant homology with other members of the superfamily, but the majority of this domain shows strong homology only with closely related proteins involved in polypeptide transport. This observation has led to the presumption that the membrane domain forms a specific transport channel for HlyA. Nevertheless, the precise role of HlyB in polypeptide (HlyA) transport is still largely obscure.

As shown in Figure 1.12, the prokaryotic ABC type ATPases involved in solute uptake (for example, maltose), which form functional dimers, are encoded

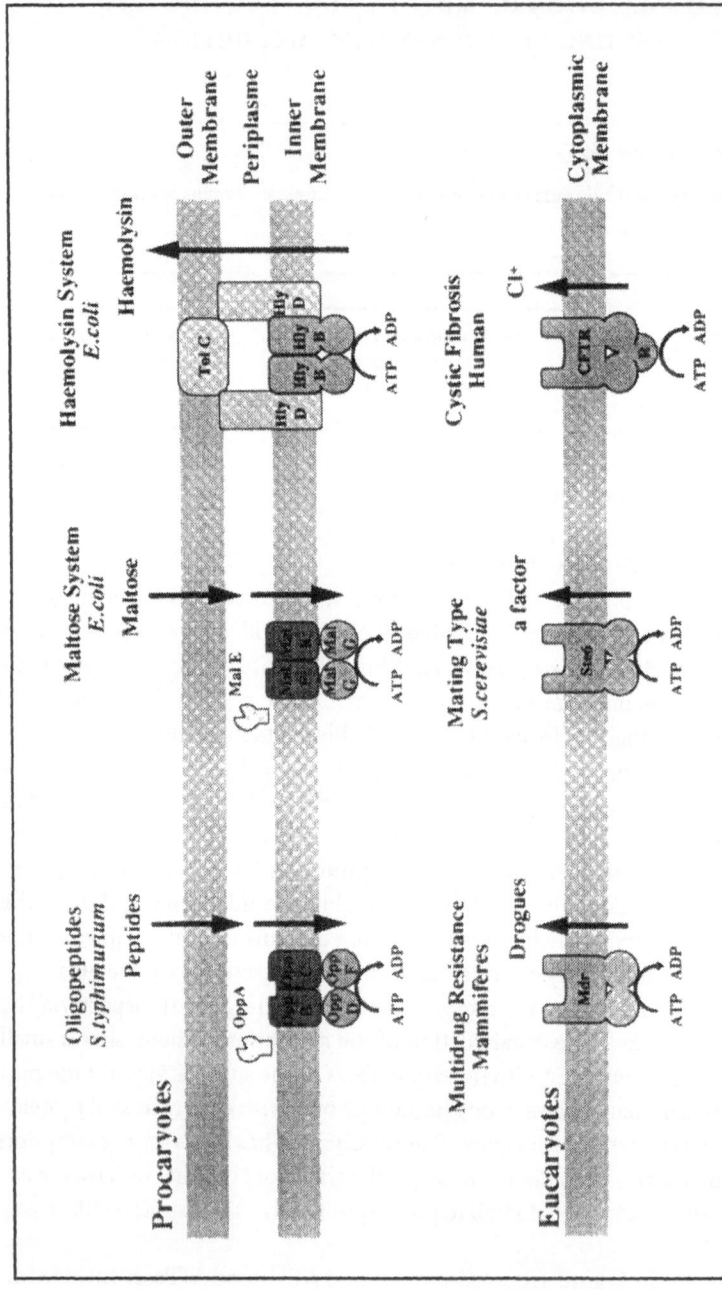

Fig. 1.12. General organization of different secretion systems using ABC proteins. *ABC ATPases* (TMD and ABC domains encoded by different genes): Opp, Mal. *ABC transporters* (TMD and ABC domains encoded by the same gene): Hly, Mdr, Ste6 and CFTR. TMD= *TransMembrane Domain*; ABC= *ATP Binding Cassette* (ATP binding domain).

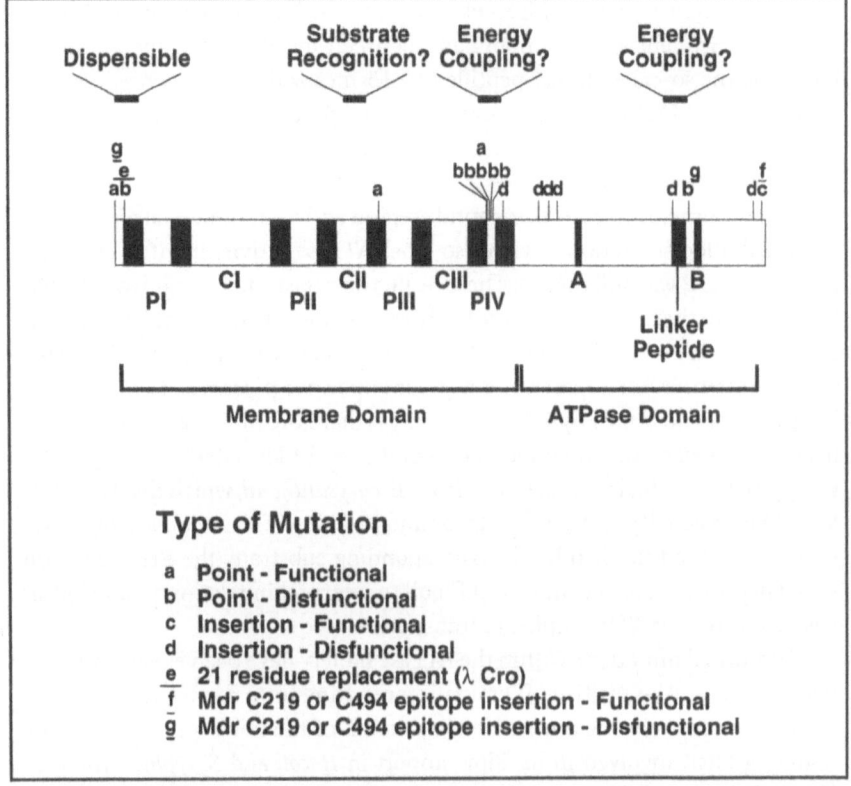

Fig. 1.13. Possible functional domains within HlyB. Bars represent transmembrane domains experimentally identified. Note that this refined model represents an amalgam of our previous β-lactamase fusion data combined with the data of the Goebel laboratory (for details see reference 23). A and B = Walker motifs. CI to CII and Pi et PIV refer to cytoplasmic and periplasmic loops respectively. Different types of mutation are indicated in the Figure. Several additional mutations (not shown, see peferences 114, 129), implicate CII in particular in interaction with HlyA since several mutations suppressing a deleted signal sequence map at this position. PIV, one of the most conserved regions in the HlyB-MDR1 family is presumed to be involved in some common function, for example energy coupling.

separately from the genes encoding the membrane (permease) domain. HlyB is the prototype ABC transporter where the ATPase is fused to the membrane domain in a single unit, in contrast to several eukaryotic ABC transporters which are tandemly duplicated. HlyB is thought to form dimers or higher multimers but specific evidence for this is not yet available.

In relation to the structure-function analysis of HlyB, at least 40 mutations in HlyB have been isolated and sequenced and these have been reviewed in detail elsewhere.[23] Therefore, only the major mutations and their effects will be briefly considered here. The HlyB C-terminal domain contains typical Walker

A and B motifs associated with ATPase activity which has indeed been detected in vitro.[123] Sequences surrounding and between the Walker A and B motifs, including the so-called linker peptide,[124] a 15-mer with the sequence in HlyA, LSGGQRQRIAIARAL, are the most conserved parts of the molecule and are virtually identical across the family and absolutely diagnostic for ABC transporters (see Fig. 1.13).

Mutations within the linker peptide region and within the Walker AB motifs, which block secretion in vivo also block ATPase activity in vitro, although in all cases ATP was still bound. These results demonstrated unequivocally that the ATPase activity of HlyB is absolutely essential for its function.[123] However, the in vitro ATPase activity of the HlyB C-terminal (separated from its membrane domain) is not apparently dependent upon the presence of either membranes or the substrate HlyA and its activity is therefore presumably uncoupled from its normal regulation under these conditions.[123] The HlyB homolog, PrtD, was partially purified in intact form from *E. chrysanthemi*, which displayed low level ATPase activity and specifically bound 8-azido-ATP. Curiously, this activity was inhibited in vitro by the corresponding substrate, the PrtB secretion signal peptide,[125] a result which is difficult to interpret in relation to possible in vivo utilization of ATP coupled to transport.

Additional mutations within the ATPase moiety have been isolated including a temperature-sensitive (Ts) mutation resulting from a Leu substitution in a conserved Pro-residue immediately upstream of the linker peptide.[126] In the homolog HisP, involved in histidine import in *E. coli* and *S. typhimurium*, an identical change renders the ATPase activity constitutive in vivo.[127,128] Another Ts mutation was identified in the relatively well-conserved extracytoplasmic loop, PIII, between TMDs 7 and 8 (Fig. 1.13), in which several other mutations leading to loss of function were identified.[126] However, the basis of the defect in these latter mutations has not been investigated, due to the lack of suitable in vivo or in vitro assays (but see below).

In attempts to study specific functions of HlyB, in particular substrate recognition, Ling and co-workers[114,129] screened for *hlyB* mutations which suppress the effect of large deletions, essentially eliminating either Helix II or Helix I within the HlyA secretion signal. These HlyB mutations, which generally increase 4-fold the low level of secretion of the signal deletion mutations, are distributed in at least four clusters, mainly in cytoplasmic regions, CI, CII and close to TMD8. Further studies will be required to establish whether any of these mutations define secretion signal recognition sites or whether suppression is obtained through other mechanisms, for example, through changes in levels of HlyB expression. Interestingly, this study also identified two mutations, asp-433-gln and asp-259-gln, which reduced the level of secretion of HlyA containing a replacement signal sequence from the close relative, leukotoxin, indicating a change in specificity of the translocator. However, it remains to be established whether this effect is expressed at the level of recognition of the secretion signal.

Other genetic studies have indicated that the first 25 residues of HlyB are dispensable for function[130] and HlyB with a His-tag at the N-terminal is biologically active (J. Young, unpublished data, this laboratory). Similarly, KpnI linkers can be inserted at the N- and C-termini without significant loss of function although such insertions at all other 6 sites abolished secretion activity.[131]

The inability to overproduce and isolate intact HlyB for in vitro reconstitution experiments has so far hampered efforts to analyze the regulation of ATPase activity in wild-type and mutant HlyB molecules. Similarly, the apparently low level of expression of *hlyB* in vivo has precluded detailed in vivo studies of the effects of mutations on hemolysin secretion. However, antibody is now available[130] and conditions for the overproduction and routine detection of HlyB in vivo have recently been established which should greatly facilitate future studies.[132]

Experiments with the Has-system, responsible for a heme acquisition protein (HasA) in *S. marcescens*, have recently provided some evidence that the HlyB homolog, HasD, participates directly and specifically in translocation of its homologous substrate, HasA.[133] HasA and the metalloprotease C (PrtC, *E. chrysanthemi*) are both secreted from *E. coli* using their own homologous translocators, however, while PrtC is secreted by both sets of translocator, secretion of HasA is only possible with its own specific translocator. This allowed Binet and Wandersman[134] to carry out mix and match experiments which showed that HasA secretion was absolutely dependent upon the presence of its own ABC transporter, HasD, but that either MFP protein could function in secretion. However, the level at which substrate discrimination by the ABC transporter, recognition or translocation, occurred, was not established in these experiments.

ANALYSIS OF HLYD FUNCTION

HlyD is absolutely essential for HlyA secretion and is a member of the so-called membrane fusion protein (MFP) family,[135] present in Gram-negative bacteria and involved in varied functions requiring close contact between inner and outer membranes. Earlier studies demonstrated that radiolabeled HlyD fractionated primarily with the cytoplasmic membrane with some indications of some molecules also fractionating with the outer membrane.[136,137] However, recent studies in this laboratory utilizing equilibrium sucrose gradient analysis, combined with HlyD antibody which is now available, have failed to confirm

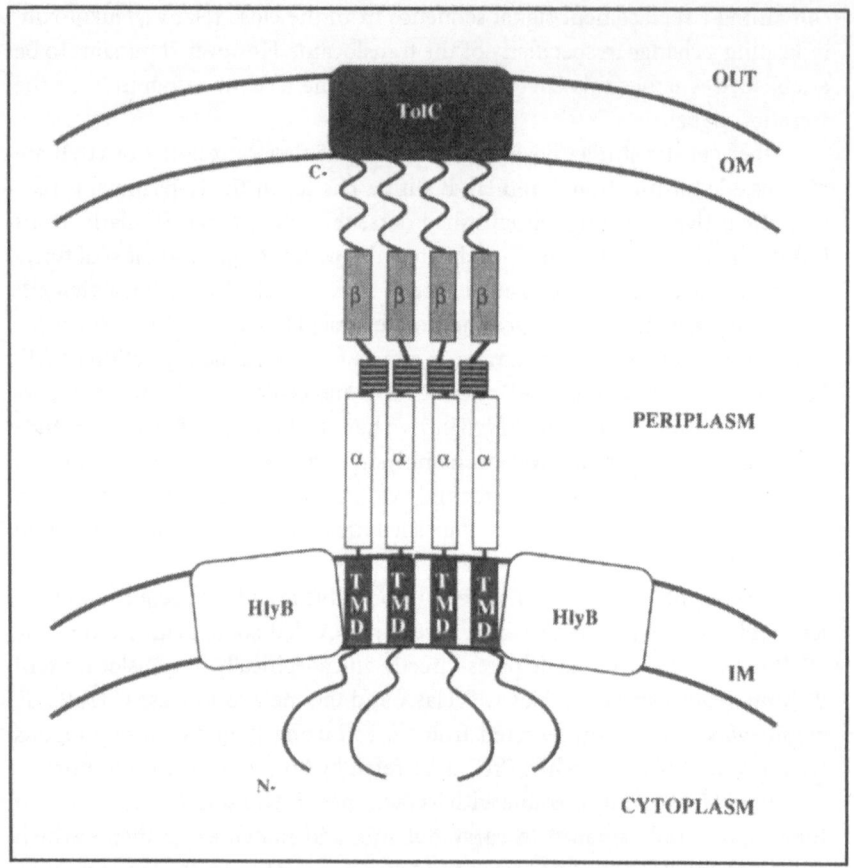

Fig. 1.14. Proposed model for the structure of the HlyD protein. HlyD is represented with its N- terminal cytoplasmic domain (*N-*), followed by the hydrophobic transmembrane domain (*TMD*) and the mainly α-helical region spanning the periplasm, with its C-terminal (*C-*) interacting with the outer membrane, possibly through TolC. *IM/OM* = Inner and Outer Membrane, respectively. α = α-helix; β = β-strand; ▦ = coiled-coil helix. Some genetic evidence suggests that HlyD is a multimer but stoichiometry has not been analyzed.

any significant cofractionation with the outer membrane and HlyD was clearly localized to the inner membrane.[32]

As presented in Figure 1.14, topological analysis of HlyD[138,139] as with other MFP proteins indicated a single transmembrane domain (TMD), with an approximately 58 residue N-terminal in the cytoplasm with a large C-terminal domain which presumably spans the periplasm. This periplasmic domain is characterized by an extensive (approximately 180 residues) α helical domain including a conserved region of coiled-coil,[140] with the C-terminal 150 residues or so predominantly of predicted β-strand. Within the close relatives of HlyD, involved in secretion of other RTX proteins, the central helical region is poorly conserved at primary sequence level. In contrast, the regions flanking the TMD and most of the β-strand domain are relatively well conserved within this MFP subgroup. In view of the similarity of the structure of HlyD and other MFP proteins, and the complete absence of periplasmic intermediates in HlyA secretion, we have proposed that HlyD forms a periplasmic chamber tightly sealed with outer membrane lipids or proteins (for example, TolC). This might be anticipated to require several molecules of HlyD and Schülein et al[137] have provided genetic evidence for HlyD multimers.

Point mutations or deletions within the C-terminal 33 residues of HlyD were previously shown to block secretion, including a substitution of the C-terminal Arg for Leu, which reduced secretion by 70%.[137] However, we have recently found that the modification of residues at the extreme C-terminus of HlyD can lead to substantial instability of HlyD in the envelope using antibody.[32] Similarly, in the absence of TolC, HlyD is apparently less stable and envelope levels are reduced. This may indicate direct interaction between HlyD and TolC but this remains to be established. However, in the mix and match experiments of Binet and Wandersman,[134] evidence was also obtained which indicated an interaction between TolC and the MFP protein, HasE, providing additional evidence that these RTX-translocators may form a continuous structure with the outer membrane.

While the C-terminal of HlyD and TolC may therefore interact, and this interaction may be essential for function and for stabilization of HlyD, removal of the N-terminal 40 amino acids of HlyD apparently does not destabilize HlyD nor prevent its assembly into the membrane.[32] Such an N-terminal deletion, however, blocks secretion of HlyA suggesting a specific role for this region of HlyD in an early stage in HlyA secretion.

In recent studies in this laboratory, random mutations were isolated in *hlyD* and 12 such mutations were sequenced. Of these, 8 mutants gave null phenotypes on blood agar and all appeared to be defective at an early stage in HlyA transport, while at least one mutant, HlyD54, appeared to accumulate HlyA in the envelope. In contrast, another mutant, D8, with a mutation in the C-terminal β-strand region (see Fig. 1.14), appeared to secrete HlyA which was incorrectly folded. As will be described in detail elsewhere, these results suggested that N-terminal and C-terminal regions of HlyD may be involved in both early and late (folding/release) steps in the secretion pathway, respectively.[141]

THE TRANSLOCATOR COMPONENT, TOLC

TolC is an outer membrane protein present in all strains of *E. coli* and implicated in the uptake of colicins, including ColEI.[142] Mutations in *tolC* are

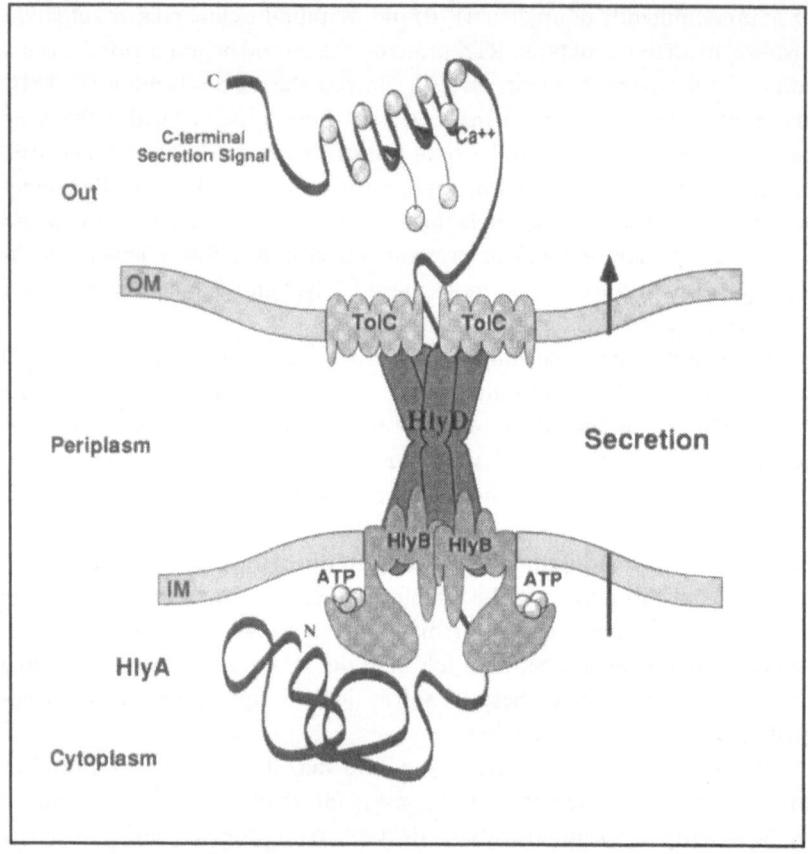

Fig. 1.15. Model proposed for the secretion of HlyA through the transenvelope translocator composed of HlyB, HlyD and TolC. IM = Inner Membrane; OM= Outer Membrane. Stoichiometry of the different translocator proteins is arbitrary.

however extremely pleiotropic, including increasing sensitivity to a wide range of drugs and detergents. This and other evidence (K. Schnaitman, personal communication) may indicate that TolC is specifically involved in LPS biogenesis. Additionally, however, TolC appears to be specifically required for hemolysin secretion.[101] Strikingly, the secretion of some other RTX proteins including PrtB is specifically dependent upon TolC homologs encoded by genes adjacent to the *hlyB, hlyD*-like translocator genes.[143] This provides strong support for the conclusion that TolC and its homologs participate directly in Type I secretion pathways, perhaps as the exit point through the outer membrane.

PERSPECTIVES AND MODEL FOR HLYA SECRETION

In 1986, we proposed a model in which the C-terminal secretion signal of HlyA was recognized by a transenvelope complex composed of HlyD and the

ABC transporter HlyB, which then directly transported the hemolysin to the external medium without any periplasmic intermediate. Subsequent studies by Wandersman and Delepelaire[101] indicated that the transenvelope transporter also included the outer membrane protein TolC. Based on the evidence reviewed above, we may now refine this model further with some additional speculation based on unpublished data. This is summarized in Figure 1.15.

We now envisage, therefore, that the 46 residue secretion signal peptide, in relatively unfolded form, recognizes and docks with elements of the translocator, including the N-terminal of HlyD and some region of the HlyB membrane domain not yet identified. In line with proposals for the activity of other ABC type ATPases such as HisP,[124] we presume that interaction with the secretion signal activates the HlyB ATPases. This in turn energizes initiation of translocation of an unfolded HlyA molecule across the cytoplasmic membrane, involving the membrane domain of HlyB acting in concert with HlyD. The membrane potential across the cytoplasmic membrane required for HlyA secretion, is also likely to play a role at this or a later stage.[144] We then envisage that an oriented translocation, HlyA-C-terminal leading, is completed through the envelope involving a multimeric chamber formed by HlyD, tightly sealed into the outer membrane through interaction with TolC. Final stages of secretion involving the secretion signal itself, the RTX-repeat region and surface Ca^{2+} may also then play crucial roles in a coordinated process that ensures that the emerging HlyA molecule folds correctly and is efficiently released from the cell surface. Recent studies from this laboratory (Blight M.A., unpublished)[141] indicated that Ca^{2+} plays an important role in the secretion-refolding process. This is reminiscent of recent studies of the secretion of some proteins from Gram-positive bacteria, which in the absence of chaperones appear to utilize Ca^{2+} as a catalyst to promote refolding.[145] This model still retains many elements of uncertainty to be resolved by future studies. Notably, the precise structure and stoichiometry of the translocator, the structure of HlyA during transit and the mechanism of folding-release to the medium are all unclear but the basic elements for the resolution of these questions appear now to be in place.

An enigma however remains, why are some secretion pathways relatively simple, while others, for example Type III, require initial translocation by the *secAY* system and then such a complex apparatus to facilitate passage across the outer membrane from the periplasm. The key to the contrast between the Type I and Type III pathways, for example, may lie in the differing requirements for the secretion of apparently fully folded molecules via the Type III pathway in contrast to probably unfolded molecules of HlyA. The apparent complexity of the Type III system may confer two advantages over the Type I mechanism. In the first instance, pullulanase secreted via Type III appears to be inherently far more stable than hemolysin and it is tempting to speculate that folding in the periplasmic compartment has considerable advantages over folding on the potentially hostile environment of the cellular surface. Secondly, the expression of the *hlyBD, tolC,* transporter complex[146] and in particular the concomitant secretion of HlyA, renders *E. coli* (including wild-type pathogenic strains), hypersensitive to several drugs in the 500 to 1500 molecular weight range;[126] (Pimenta A, Jaouadi N, Holland IB, unpublished data). These results clearly imply that secretion of HlyA compromises the integrity of the surface envelope, providing a "channel," accessible to a number of drugs normally excluded by the outer membrane. Restriction of hemolysin secretion to a narrow window during late exponential phase[107] may, therefore, be an essential adaptation to minimize this apparent penalty associated with hemolysin secretion.

REFERENCES
1. Pugsley AP. ed. Protein Targeting. San Diego: Academic Press, 1989.
2. Blight MA, Holland IB. Structure and function of haemolysin B, P-glycoprotein and other members of a novel family of membrane translocators. Mol Microbiol 1990; 4: 873-880.
3. Salmond GPC, Reeves PJ. Membrane traffic wardens and protein secretion in Gram-negative bacteria. Trends Cell Biol 1993; 18: 7-12.
4. Genin S, Boucher CA. A superfamily of proteins involved in different secretion pathways in Gram-negative bacteria: modular structure and specificity of the N-terminal domain. Mol Gen Genet 1994; 243: 112-118.
5. Blight MA, Chervaux C, Holland IB. Protein secretion in *Escherichia coli.* Curr Opinion Biotech 1994; 5: 468-474.
5a. Lauer P, Albertson NH, Koomeg M. Conservation of genes encoding components of a type IV pilus assembly/two-step protein export pathway in N. gonorrhoeae. Mol Microbiol 1993; 8: 357-368.
6. Pugsley AP, Kornacker MG, Poquet I. The general protein-export pathway is directly required for extracellular pullulanase secretion in *Escherichia coli* K12. Mol Microbiol 1991; 5: 343-352.
7. d'Enfert C, Ryter A, Pugsley AP. Cloning and expression in *Escherichia coli* of the *Klebsiella pneumoniae* genes for production, surface localization and secretion of the lipoprotein pullulanase. EMBO J 1987; 6: 3531-3538.
8. Pugsley AP, d'Enfert C, Reyss I et al. Genetics of extracellular protein secretion by Gram-negative bacteria. Ann Rev Genet 1990; 24: 67-90.
9. He SY, Lindeberg M, Chatterjee AK et al. Cloned *Erwinia chrysanthemi out* genes enable *Escherichia coli* to selectively secrete a diverse family of heterologous proteins to its milieu. Proc Natl Acad Sci USA 1991; 88: 1079-1083.
10. Reeves PJ, Whitcombe D, Wharam S et al. Molecular cloning and characterization of 13 out genes from *Erwinia carotovora* subspecies *carotovora*: genes encoding members of a general secretion pathway (GSP) widespread in Gram-negative bacteria. Mol Microbiol 1993; 8: 443-456.

11. Dums F, Dow JM, Daniels MJ. Structural characterization of protein secretion genes of the bacterial phytopathogen *Xanthomonas campestris* pathovar *campestris*: relatedness to secretion systems of other Gram-negative bacteria. Mol Gen Genet 1991; 229: 357-64.

12. Filloux A, Bally M, Ball G et al. Protein secretion in Gram-negative bacteria: transport across the outer membrane involves common mechanisms in different bacteria. EMBO J 1990; 9: 4323-4329.

13. Filloux A, Bally M, Murgier M et al. Cloning of *xcp* genes located at the 55 min region of the chromosome and involved in protein secretion in *Pseudomonas aeruginosa*. Mol Microbiol 1989; 3: 261-265.

14. Howard SP, Critch J, Bedi A. Isolation and analysis of eight exe genes and their involvement in extracellular protein secretion and outer membrane assembly in Aeromonas hydrophila. J Bacteriol 1993; 175: 6695-6703.

15. Pugsley AP. The complete general secretory pathway in Gram-negative bacteria. Microbiol Rev 1993; 57: 50-108.

16. Hobbs M, Mattick JS. Common components in the assembly of type 4 fimbriae, DNA transfer systems, filamentous phage and protein-secretion apparatus: a general system for the formation of surface-associated protein complexes. Mol Microbiol 1993; 10: 233-243.

17. Possot O, d'Enfert C, Reyss I et al. Pullulanase secretion in *Escherichia coli* K-12 requires a cytoplasmic protein and a putative polytopic cytoplasmic membrane protein. Mol Microbiol 1992; 6: 95-105.

18. Possot O, Pugsley AP. Molecular characterization of PulE, a protein required for pullulanase secretion. Mol Microbiol 1994; 12: 287-299.

19. Walker JE, Saraste M, Runswick MJ et al. Distantly related sequences in the a- and b-subunits of ATP synthase, myosin, kinase and other ATP-requiring enzymes and a common nucleotide binding fold. EMBO J 1982; 1: 945-951.

20. Nunn DN, Bergman S, Lory S. Products of three accessory genes, *pilB*, *pilC* and *pilD* are required for biogenesis of *Pseudomonas aeruginosa* pili. J Bacteriol 1990; 172: 2911-2919.

21. Sandkvist M, Bagdasarian M, Howard SP et al. Interaction between the autokinase EpsE and EpsL in the cytoplasmic membrane is required for extracellular secretion in *Vibrio cholerae*. EMBO J 1995; 14: 1664-1673.

22. Ames GF, Mimura CS, Shyalama V. Bacterial periplasmic permeases belong to a family of transport proteins operating from *Escherichia coli* to human: traffic ATPases. FEMS Microbiol Rev 1990; 6: 429-446.

23. Holland IB, Blight MA. Structure and function of HlyB, the ABC transporter essential for haemolysin secretion from *Escherichia coli*. In: The Handbook of Biological Physics "Transport Processes in Eukaryotic and Prokaryotic Organisms." Hoff AJ. Series Ed. North Holland Press. 2: 111-135.

24. Pugsley AP, Possot O. The general secretory pathway of *Klebsiella oxytoca*: no evidence for relocalization or assembly of pilin-like PulG protein into a multiprotein complex. Mol Microbiol 1993; 10: 665-674.

25. d'Enfert C, Pugsley AP. *Klebsiella pneumoniae pulS* gene encodes an outer membrane lipoprotein required for pullulanase secretion. J Bacteriol 1989; 171: 2673-3679.
26. Kazmierczak BI, Mielke DL, Russel M et al. pIV, a filamentous phage protein that mediates phage export across the bacterial cell envelope, forms a multimer. J Mol Biol 1994; 238: 187-198.
27. Hardie KR, Lory S, Pugsley T. Insertion of an outer membrane protein in *Escherichia coli* requires a chaperone-like protein. EMBO J 1996; 15: 978-988.
28. Nunn D, Lory S. Product of the *Pseudomonas aeruginosae* gene *pilD* is a prepilin peptidase. Proc Natl Acad Sci USA 1991; 88: 3281-3285.
29. Pugsley AP, Dupuy B. An enzyme with type IV prepilin peptidase activity is required to process components of the general extracellular protein secretion pathway of *Klebsiella oxytoca*. Mol Microbiol 1992; 6: 751-760.
30. Strom MS, Nunn DN, Lory S. A single bifunctional enzyme, PilD, catalyzes cleavage and N-methylation of proteins belonging to the type IV pilin family. Proc Nat Acad Sci USA 1993; 90: 2404-2408.
31. Nunn DN, Lory S. Cleavage, methylation, and localization of the *Pseudomonas aeruginosa* export proteins XcpT, -U, -V, and -W. J Bacteriol 1993; 175: 4375-4382.
32. Pimenta AL, Holland IB, Blight MA. Over-expression of the *Escherichia coli* HlyD polypeptide, production of polyclonal antibody and in vivo HlyB and TolC dependent localization and assembly. (In preparation).
33. Py B, Chippaux M, Barras F. Mutagenesis of cellulase EGZ for studying the general protein secretory pathway in *Erwinia chrysanthemi*. Mol Microbiol 1993; 7: 785-793.
34. Sauvonnet N, Poquet I, Pugsley AP. Extracellular secretion of pullulanase is unaffected by minor sequence changes but is usually prevented by adding reporter proteins to its N- or C-terminal end. J Bacteriol 1995; 177: 5238-5246.
35. Lu H-M, Lory S. A specific targetting domain in mature exotoxin A is required for its extracellular secretion from *Pseudomonas aeruginosa*. EMBO J 1996; 15: 429-436.
35a. Sauvonnet N, Pugsley AP. Identification of two regions of *K. oxytoca* pullulanase that together are capable of promoting β-lactamase secretion by the General Secretory Pathway. Mol Microbiol 1996; 22: 1-7.
36. Hirst TR, Holmgren J. Conformation of protein secreted across bacterial outer membranes: a study of enterotoxin translocation from *Vibrio cholerae*. Proc Natl Acad Sci USA 1987; 84: 7418-7422.
37. Hofstra H, Witholt B. Kinetics of synthesis, processing, and membrane transport of heat-labile enterotoxin, a periplasmic protein in *Escherichia coli*. J Biol Chem 1984; 259: 15182-7.
38. Hofstra H, Witholt B. Heat-labile enterotoxin in *Escherichia coli*. Kinetics of association of subunits into periplasmic holotoxin. J Biol Chem 1985; 260: 16037-44.

39. Hirst TR, Holmgren J. Transient entry of enterotoxin subunit into the periplasm occurs during secretion from *Vibrio cholerae.* J Bacteriol 1987; 169: 1037-1045.

40. Pugsley AP. Translocation of a folded protein across the outer membrane in *E.coli.* Proc Natl Acad Sci USA 1992; 89: 12058-12062.

41. Bortoli-German I, Brun E, Py B et al. Periplasmic disulphide bond formation is essential for cellulase secretion by the plant pathogen *Erwinia chrysanthemi.* Mol Microbiol 1994; 11: 545-553.

42. Hayano T, Takahashi N, Kato S et al. Two distinct forms of peptidylprolyl-cis-*trans*-isomerase are expressed separately in periplasmic and cytoplasmic compartments of *Escherichia coli* cells. Biochemistry 1991; 30: 3041-3048.

43. Chen R, Henning U. A periplasmic protein (Skp) of *Escherichia coli* selectively binds a class of outer membrane proteins. Mol Microbiol 1996; 19: 1287-1294.

44. Pohlner J, Halter R, Beyreuther K et al. Gene structure and extracellular secretion of *Neisseria gonorrhoeae* IgA protease. Nature 1987; 325: 458-462.

45. Klauser T, Pohlner J, Meyer TF. The secretion pathway of IgA protease-type proteins in Gram-negative bacteria. BioEssays 1993; 15: 799-805.

46. Pouisen K, Hjorth JP, Killian M. Limited diversity of immunoglobulin A1 protease gene (*iga*) among *Haemophilus influenzae* sertype b strains. Infect Immun 1988; 56: 987-992.

47. Klauser T, Kramer J, Otzelberger K et al. Characterization of the *Neisseria* IgA β-core. The essential unit for outer membrane targeting and extracellular protein secretion. J Mol Biol 1993; 234: 579-593.

48. Miyazaki H, Yanagida N, Horinouchi S et al. Characterization of the precursor of *Serratia marcescens* serine protease and C-terminal processing of the precursor during its excretion through the outer membrane of *Escherichia coli.* J Bacteriol 1989; 171: 6566-6572.

49. Suzuki T, Lett M-C, Sasakawa C. Extracellular transport of VirG protein in *Shigella.* J Biol Chem 1995; 270: 30874-30880.

50. Schmitt W, Haas R. Genetic analysis of *Helicobacter pylori* vacuolating cytotoxin: structural similarities with the IgA protease type of exported protein. Mol Microbiol 1994; 12: 307-319.

51. Randall LL, Hardy SJS. Correlation of competence for export with lack of tertiary structure of the mature species: a study in vivo of maltose-binding protein in *E. coli.* Cell 1986; 46: 921-928.

52. Ohnishi Y, Nishiyama M, Horinouchi S et al. Involvement of the COOH-terminal pro-sequence of *Serratia marcescens* serine protease in the folding of the mature enzyme. J Biol Chem 1994; 269: 32800-32806.

53. Hultgren SJ, Abraham S, Caparon M et al. Pilus and nonpilus bacterial adhesins: assembly and function in cell recognition. Cell 1993; 73: 887-901.

54. Jacob-Dubuisson F, Heuser J, Dodson K et al. Initiation of assembly

and association of the structural elements of a bacterial pilus depend on two specialized tip proteins. EMBO J 1993; 12: 837-47.

55. Tennent JM, Lindberg F, Normark S. Integrity of *Escherichia coli* P pili during biogenesis: properties and role of PapJ. Mol Microbiol 1990; 4: 747-58.

56. Hultgren SJ, Normark S, Abraham SN. Chaperone-assisted assembly and molecular architecture of adhesive pili. Ann Rev Microbiol 1991; 45: 383-415.

57. Backer D, Vader CEM, Roosendaal B et al. Structure and function of periplasmic chaperone-like proteins involved in the biosynthesis of K88 and K99 fimbriae in enterotoxigenic *Escherichia coli*. Mol Microbiol 1991; 5: 875-886.

58. Hultgren S, Lindberg F, Magnusson G et al. The PapG adhesin of uropathogenic *Escherichia coli* contains separate regions for receptor binding and for incorporation into the pilus. J Bacteriol 1989; 86: 4357-4361.

59. Jacob-Dubuisson F, Striker R, Hultgren SJ. Chaperone-assisted self-assembly of pili independent of cellular energy. J Biol Chem 1994; 269: 12447-12455.

60. Lauer P, Albertson NH, Koomey M. Conservation of genes encoding components of a type IV pilus assembly/two step protein export pathway in *Neisseria gonorrhoeae*. Mol Microbiol 1993; 8: 357-368.

61. Braun V, Neuss B, Raun Y et al. Identification of the *Serratia marcescens* haemolysin determinant by cloning into *Escherichia coli*. J Bacteriol 1987; 169: 2113-2120.

62. Uphoff TS, Welch RA. Nucleotide sequencing of the *Proteus mirabilis* calcium-independent haemolysin genes (*hpmA* and *hpmB*) reveals sequence similarity with the *Serratia marcescens* haemolysin genes (*shlA* and *shlB*). J Bacteriol 1990; 172: 1206-1216.

63. Jacob-Dubuisson F, Buisine C, Mielcarek N et al. Amino-terminal maturation of the *Bordetella pertussis* filamentous haemagglutinin. Mol Microbiol 1996; 19: 65-78.

64. Foreman DT, Martinez Y, Coombs G et al. TolC and DsbA are needed for secretion of STB, a heat-stable enterotoxin of *Escherichia coli*. Mol Microbiol 1995; 18: 237-245.

65. Braun V, Ondraczek R, Hobbie S. Activation and secretion of *Serratia* haemolysin. Int J Med Microbiol Virol Parasitol Infect Dis 1993; 278: 306-315.

66. Braun V, Herrmann C. Evolutionary relationship of uptake systems for biopolymers in *Escherichia coli*: cross-complementation between the TonB-ExbB-ExbD and the TolA-TolQ-TolR proteins. Mol Microbiol 1993; 8: 261-268.

67. Brubaker RR. Factors promoting acute and chronic diseases caused by *Yersiniae*. Clin Microbiol Rev 1991; 4: 309-324.

68. Forsberg Å, Bölin I, Norlander L et al. Molecular cloning and expression of calcium-regulated, plasmid encoded proteins of *Y. pseudotuberculosis*. Microb Pathogen 1987; 2: 123-137.

69. Cornelis GR, Biot T, Rouvroit CLd et al. The *Yersinia* yop regulon. Mol Microbiol 1989; 3: 1455-1459.
70. Michiels T, Vanooteghem JC, Rouvroit CLd et al. Analysis of *virC*, an operon involved in the secretion of Yop proteins by *Yersinia enterocolitica*. J Bacteriol 1991; 173: 4994-5009.
71. Michiels T, Wattiau P, Brasseur R et al. Secretion of Yop proteins by *Yersiniae*. Infect Immun 1990; 58: 2840-2849.
72. Plano GV, Barve SS, Straley SC. LcrD, a membrane-bound regulator of the *Yersinia pestis* low-calcium response. J Bacteriol 1991; 173: 7293-7303.
73. Woestyn S, Allaoui A, Wattiau P et al. YscN, the putative energizer of the *Yersinia* Yop secretion machinery. J Bacteriol 1994; 176: 1561-1569.
74. Plano GV, Straley SC. Mutations in yscC, yscD, and yscG prevent high-level expression and secretion of V antigen and Yops in *Yersinia pestis*. J Bacteriol 1995; 177: 3843-3854.
75. Michiels T, Cornelis GR. Secretion of hybrid proteins by the *Yersinia* Yop export system. J Bacteriol 1991; 173: 1677-1685.
76. Wattiau P, Cornelis GR. SycE, a chaperone-like protein of *Yersinia enterocolitica* involved in the secretion of YopE. Mol Microbiol 1993; 8: 123-131.
77. Wattiau P, Bernier B, Deslée P et al. Individual chaperones required for Yop secretion. Proc Natl Acad Sci USA 1994; 91: 10493-10497.
78. Ménard R, Sansonetti P, Parsot C et al. Extracellular association and cytoplasmic partitioning of the IpaB and IpaC invasins of *S. flexneri*. Cell 1994; 79: 515-525.
79. Wattiau P, Woestyn S, Cornelis GR. Customized secretion chaperones in pathogenic bacteria. Molec Microbiol 1996; 20: 255-262.
80. Van Gijsegem F, Gough C, Zischek C et al. The hrp gene locus of *Pseudomonas solanacearum*, which controls the production of a type III secretion system, encodes eight proteins related to components of the bacterial flagellar biogenesis complex. Mol Microbiol 1995; 15: 1095-1114.
81. Allaoui A, Sansonetti PJ, Parsot C. MxiJ, a lipoprotein involved in secretion of *Shigella* Ipa invasins, is homologous to YscJ, a secretion factor of the *Yersinia* Yop proteins. J Bacteriol 1992; 174: 7661-7669.
82. Allaoui A, Sansonetti PJ, Parsot C. MxiD, an outer membrane protein necessary for the secretion of the *Shigella flexneri* Ipa invasins. Mol Microbiol 1993; 7: 59-68.
83. Galan JE, Ginocchio C, Costeas P. Molecular and functional characterization of the *Salmonella invasion* gene invA: homology of InvA to members of a new protein family. J Bacteriol 1992; 174: 4338-4349.
84. Jarvis KG, Giron JA, Jerse AE et al. Enteropathogenic *Escherichia coli* contains a putative type III secretion system necessary for the export of proteins involved in attaching and effacing lesion formation. Proc Natl Acad Sci USA 1995; 92: 7996-8000.
85. Fenselau S, Balbo I, Bonas U. Determinants of pathogenicity in

Xanthomonas campestris pv. vesicatoria are related to proteins involved in secretion in bacterial pathogens of animals. Mol Plant-Microbe Interact 1992; 5: 390-396.

86. Xiao Y, Lu Y, Heu S et al. Organization and environmental regulation of the *Pseudomonas syringae* pv. syringae 61 *hrp* cluster. J Bacteriol 1992; 174: 1734-1741.

87. Galan JE. Molecular genetic bases of *Salmonella* entry into host cells. Molec Microbiol 1996; 20: 263-271.

88. Ménard R, Sansonetti P, Parsot C. The secretion of the *S. flexneri* Ipa invasins is induced by epithelial cells and controlled by Ipab and IpaD. EMBO J 1994; 13: 5293-5302.

89. Rosqvist R, Hakansson S, Forsberg A et al. Functional conservation of the secretion and translocation machinery for virulence proteins of yersiniae, salmonellae and shigellae. EMBO J 1995; 14: 4187-4195.

90. Barinaga M. A shared strategy for virulence. Science 1996; 272: 1261-1263.

91. Russel M. Moving through the membrane with filamentous phages. Trends Microbiol 1995; 3: 223-228.

92. Guy-Caffey J, Rapoza MP, Jolley KA et al. Membrane localization and topology of a viral assembly protein. J Bacteriol 1992; 174: 2460-2465.

93. Guy-Caffey J, Webster RE. The membrane domain of a bacteriophage assembly protein. Membrane insertion and growth inhibition. J Biol Chem 1993; 268: 5496-54503.

94. Guy-Caffey J, Webster RE. The membrane domain of a bacteriophage assembly protein. Transmembrane- directed proteolysis of a membrane-spanning fusion protein. J Biol Chem 1993; 268: 5488-5495.

95. Brissette JL, Russel M. Secretion and membrane integration of a filamentous phage-encoded morphogenetic protein. J Mol Biol 1990; 211: 565-580.

96. Russel M, Kazmierczak B. Analysis of the structure and subcellular location of filamentous phage pIV. J Bacteriol 1993; 175: 3998-4007.

97. Ludwig A, Goebel W. Genetic determinants of cytolytic toxins from Gram-negative bacteria. ed. Sourcebook of Bacterial Proteins. Academic Press Ltd., 1991; 117-145.

98. Kenny B, Finlay BB. Protein secretion by enteropathogenic *Escherichia coli* is essential for transducing signals to epithelial cells. Proc Natl Acad Sci USA 1995; 92: 7991-7995.

99. Stanley P, Packman LC, Koronakis V et al. Fatty acylation of two internal lysine residues required for the toxic activity of *Escherichia coli* haemolysin. Science 1994; 266: 1992-1996.

100. Nicaud J-M, Mackman N, Gray L et al. Characterization of HlyC and mechanism of activation and secretion of haemolysin from *E. coli* 2001. FEBS Lett 1985; 187: 339-344.

101. Wandersman C, Delepelaire P. TolC, an *Escherichia coli* outer membrane protein required for haemolysin secretion. Proc Natl Acad Sci USA 1990; 87: 4776-4780.

102. Mackman N, Nicaud J-M, Gray L et al. Secretion of haemolysin by *Escherichia coli*. Curr Top Microbiol Immunol 1986; 125: 159-181.

103. Boehm DF, Welch RA, Snyder IS. Domains of *Escherichia coli* haemolysin (HlyA) involved in binding of calcium and erythrocyte membranes. Infect Immun 1990; 58: 1959-1964.

104. Jarchau T, Chakraburty T, Garcia F et al. Selection for transport competence of C-terminal polypeptides derived from *Escherichia coli* haemolysin: the shortest peptide capable of autonomous HlyB/HlyD-dependent secretion comprises the C-terminal 62 amino acids of HlyA. Mol Gen Genet 1994; 245: 53-60.

105. Ghigo J-M, Wandersman C. A carboxyl-terminal four-amino acid motif is required for secretion of the metalloprotease PrtG through the *Erwinia chrysanthemi* protease secretion pathway. J Biol Chem 1994; 269: 1-7.

106. Kenny B, Haigh R, Holland IB. Analysis of the haemolysin transport process through the secretion from *Escherichia coli* of PCM, CAT or β-galactosidase fused to the Hly C- terminal signal domain. Mol Microbiol 1991; 5: 2557-2568.

107. Holland IB, Kenny B, Blight M. Haemolysin secretion from *E.coli*. Biochimie 1990; 72: 131-141.

108. Létoffé S, Wandersman C. Secretion of Cya-PrtB and HlyA-PrtB fusion proteins in *Escherichia coli*: involvement of glycine-rich repeat domains of *Erwinia chrysanthemi* protease B. J Bacteriol 1992; 174: 4920-4927.

109. Zhang F, Greig DI, Ling V. Functional replacement of the haemolysin A transport signal by a different primary sequence. Proc Natl Acad Sci USA 1993; 90: 4211-4215.

110. Sebo P, Ladant D. Repeat sequences in the *Bordetella pertussis* adenylate cyclase toxin can be recognized as alternative carboxyproximal secretion signals by the *Escherichia coli* α-haemolysin translocator. Mol Microbiol 1993; 9: 999-1009.

111. Stanley P, Koronakis V, Hughes C. Mutational analysis supports a role for multiple structural features in the C-terminal secretion signal of *Escherichia coli* haemolysin. Mol Microbiol 1991; 5: 2391-2403.

112. Kenny B, Taylor S, Holland IB. Identification of individual amino acids required for secretion within the haemolysin (HlyA) C-terminal targeting region. Mol Microbiol 1992; 6: 1477-1489.

113. Chervaux C, Holland IB. Random and directed mutagenesis to elucidate the functional importance of helix II and F-989 in the C-terminal secretion signal of *Escherichia coli* haemolysin. J Bacteriol 1996; 178: 1232-1236.

114. Zhang F, Sheps JA, Ling V. Complementation of transport-deficient mutants of *Escherichia coli* α-haemolysin by second-site mutations in the transporter haemolysin B. J Biol Chem 1993; 268: 19889-19895.

115. Kenny B, Chervaux C, Holland IB. Evidence that residues -15 to -46 of the haemolysin secretion signal are involved in early steps in secretion, leading to recognition of the translocator. Mol Microbiol 1994; 11: 99-109.

116. Chervaux C, Sauvonnet N, A. LC et al. Secretion of active β-lactamase to the medium mediated by the _Escherichia coli_ haemolysin transport pathway. Mol Gen Genet 1995; 249: 237-245.
117. Baumann U, Wu S, Flaherty KM et al. Three-dimensional structure of the alkaline protease of _P.aeruginosa_: a two-domain protein with a calcium binding parallel beta roll motif. EMBO J 1993; 12: 3357-3364.
118. Baumann U. Crystal structure of the 50 kDa metalloprotease from _Serratia marcescens_. J Mol Biol 1994; 242: 244-251.
119. Wolff N, Ghigo JM, Délépelaire P et al. C-terminal secretion signal of an _Erwinia chrysanthemi_ protease secreted by a signal peptide-independent pathway: proton NMR and CD conformational studies in membrane mimetic environements. Biochemistry 1994; 33: 6792-6801.
120. Yin Y, Zhang F, Ling V et al. Structural analysis and comparison of the C-terminal transport signal domains of haemolysin A and leukotoxin A. FEBS Lett 1995; 366: 1-5.
121. Zhang F, Yin Y, Arrowsmith CH et al. Secretion and circular dichroism analysis of the C-terminal signal peptides of HlyA and LktA. Biochemistry 1995; 34: 4193-4201.
122. Felmlee T, Pellett S, Welch RA. Nucleotide sequence of an _Escherichia coli_ chromosomal haemolysin. J Bacteriol 1985; 163: 94-105.
123. Koronakis V, Hughes C, Koronakis E. ATPase activity and ATP/ADP-induced conformational change in the soluble domain of the bacterial protein translocator HlyB. Mol Microbiol 1993; 8: 1163-1175.
124. Ames GF-L, Lecar H. ATP dependent bacterial transporters and cystic fibrosis: analogy between channels and transporters. FASEB 1992; 6: 2660-2666.
125. Délépelaire P. PrtD, the integral membrane ATP-ase-binding cassette component of the _Erwinia chrysanthemi_ metalloprotease secretion system, exhibits a secretion signal-regulated ATPase activity. J Biol Chem 1994; 269: 27952-27957.
126. Blight MA, Pimenta AL, Lazzaroni JC et al. Identification and preliminary characterization of temperature- sensitive mutations affecting HlyB, the translocator required for the secretion of haemolysin (HlyA) from _Escherichia coli_. Mol Gen Genet 1994; 245: 431-440.
127. Shyamala V, Baichwal V, Beall E et al. Structure-function analysis of the histidine permease and comparison with cystic fibrosis mutations. J Biol Chem 1991; 266: 18714-18719.
128. Petronilli V, Ames GF-L. Binding protein-independent histidine permease mutants: uncoupling of ATP hydrolysis from transmembrane signaling. J Biol Chem 1991; 266: 16293-16296.
129. Sheps JA, Cheung I, Ling V. Hemolysin transport in _Escherichia coli_. Point mutants in HlyB compensate for a deletion in the predicted amphiphilic helix region of the HlyA signal. J Biol Chem 1995; 270: 14829-14834.
130. Blight MA, Menichi B, Holland IB. Evidence for post-transcriptional regulation of the synthesis of the _Escherichia coli_ HlyB haemolysin

translocator and production of polyclonal anti-HlyB antibody. Mol Gen Genet 1995; 247: 73-85.

131. Juranka P, Zhang F, Kulpa J et al. Characterization of the haemolysin transporter, HlyB, using an epitope insertion. J Biol Chem 1992; 267: 3764-3770.

132. Young J, Holland IB, Blight MA. Solubilisation and antibody detection of the HlyB ABC-transporter by a modified Western blot technique. (In preparation).

133. Létoffé S, J.-M.Ghigo, Wandersman C. Secretion of the *Serratia marcescens* HasA protein by an ABC transporter. J Bacteriol 1994; 176: 5372-5377.

134. Binet R, Wandersman C. Protein secretion by hybrid bacterial ABC transporters: specific functions of the membrane ATPase and the membrane fusion protein. EMBO J 1995; 14: 2298-2306.

135. Dinh T, Paulsen IT, Saier-JR MH. A family of extracytoplasmic proteins that allow transport of large molecules across the outer membranes of Gram-negative bacteria. J Bacteriol 1994; 176: 3825-3831.

136. Mackman N, Nicaud J-M, Gray L et al. Identification of polypeptides required for the export of haemolysin 2001 from *E. coli*. Mol Gen Genet 1985; 201: 529-536.

137. Schülein R, Gentschev I, Schlor S et al. Identification and characterization of two functional domains of the haemolysin translocator protein HlyD. Mol Gen Genet 1994; 245: 203-211.

138. Wang RC, Séror SJ, Blight M et al. Analysis of the membrane organization of an *Escherichia coli* protein translocator, HlyB, a member of a large family of prokaryote and eukaryote surface transport proteins. J Mol Biol 1991; 217: 441-454.

139. Schülein R, Gentschev I, Mollenkopf HJ et al. A topological model for the haemolysin translocator protein HlyD. Mol Gen Genet 1992; 234: 155-163.

140. Pimenta AL, Blight MA, Clarke D et al. The Gram-Negative Cell Envelope "Springs" to Life: Coiled-Coil Trans-Envelope Proteins. Mol Microbiol 1996; 19:

141. Pimenta AL, Jamieson L, Holland IB. Analysis of mutants of HlyD, an *Escherichia coli* transenvelope protein, indicates a role in early and late steps in secretion of haemolysin, including a final folding stage. Submitted 1996;

142. Cramer WA, Heymann JB, Schendel SL et al. Structure-function of the channel forming colicins. Ann Rev Biophys Biomol Struct 1995; 24: 611-641.

143. Létoffé S, Délépelaire P, Wandersman C. Protease secretion by *Erwinia chrysanthemi*: the specific secretion functions are analogous to those of *Escherichia coli* α-haemolysin. EMBO J 1990; 9: 1375-1382.

144. Koronakis V, Hughes C, Koronakis E. Energetically distinct early and late stages of HlyB/HlyD-dependent secretion across both *Escherichia coli* membranes. EMBO J 1991; 10: 3263-3272.

145. Petit-Glatron M-F, Grajcar L, Munz A et al. The contribution of the cell wall to a transmembrane calcium gradient could play a key role in *Bacillus subtilis* protein secretion. Mol Microbiol 1993; 9: 1097-1106.
146. Wandersman C, Létoffé S. The study of vancomycin resistance mutants reveals a role for lipopolysaccharide in the secretion of *Escherichia coli* a-haemolysin and *Erwinia chrysanthemi* proteases. Mol Microbiol 1993; 7: 141-150.

CHAPTER 2

UNUSUAL PROTEIN SECRETION AND TRANSLOCATION PATHWAYS IN YEAST: IMPLICATION OF ABC TRANSPORTERS

Karl Kuchler and Ralf Egner

I. INTRODUCTION

In eukaryotic cells, the molecular machinery responsible for protein translocation across the endoplasmic reticulum (ER) membrane is well known. It comprises the cytoplasmic signal recognition particle, membrane receptors, chaperones and processing enzymes in the ER lumen.[1] Peptides and proteins destined for secretion or which have to be transported to other intracellular organelles such as the lysosome are usually synthesized as larger precursors containing a cleavable signal peptide at or near the N-terminus of the polypeptide.[2] Once properly assembled in the ER lumen, proteins are delivered to their final destinations through the exocytic pathway.[3] Protein sorting and targeting is accomplished by vectorial generation of highly specialized transport vesicles in the ER and Golgi of both yeast[4] and mammalian cells.[3] The yeast Golgi complex also represents the cellular intersection where the proper sorting of plasma membrane proteins from vacuolar polypeptides is achieved.[5,6]

There are, however, notable exceptions to this general scenario of protein secretion and trafficking both in mammalian cells (see chapters 3 and 6) and in yeast.[7] For instance, there are proteins and peptides that lack a typical hydrophobic signal peptide but are nevertheless secreted by mechanisms which apparently do not require a functional secretory pathway. A well-described example is the mating pheromone **a**-factor, which is released from budding yeast.[8,9] It has been established that the cellular export of this peptide hormone is mediated by Ste6,[10,11] a prominent member the superfamily of ATP binding cassette (ABC) transporters.

Unusual Secretory Pathways: From Bacteria to Man, edited by Karl Kuchler, Anna Rubartelli and Barry Holland. © 1997 R.G. Landes Company.

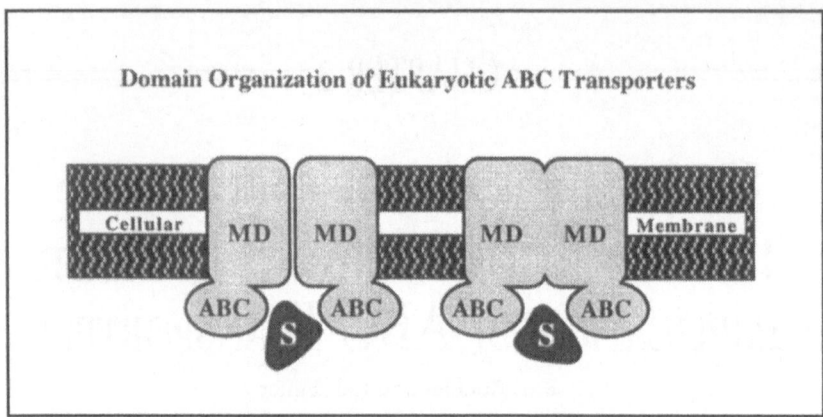

Fig. 2.1. Domain organization of eukaryotic ABC transporters. The structural organization of all eukaryotic ABC proteins is very similar, with a membrane domain (MD) and the most highly conserved ATP binding cassette (ABC). Substrate (S) recognition and binding usually occurs at the same side where the ABC domain is located, although some substrates may also be recognized and bound from within the lipid bilayer.

ABC transporters comprise a novel family of membrane transport proteins that are found operating from bacteria to man.[12] The hallmark characteristics of most ABC proteins include the presence of two highly conserved domains for ATP binding (ABC), and two membrane domains each containing usually six membrane-spanning α-helices (TMS). These four domains normally form a $(TMS_6-ABC)_2$ or $(ABC-TMS_6)_2$ configuration, but half-size transporters with a TMS_6-ABC or $ABC-TMS_6$ topology as well as other topologies are also frequently found (Fig. 2.1; Table 2.1).[7,12] Although the predicted structural organization has been highly conserved throughout evolution (Fig. 2.1), most ABC proteins appear to be of rather limited substrate specificity. Thus, ABC transporters are implicated in a remarkable variety of transport processes, including the transmembrane transport of ions, heavy metals, carbohydrates, anticancer drugs, amino acids, steroids, glucocorticoids, mycotoxins, antibiotics, pigments and, of course, peptides and proteins.[7,8,12-14] The mechanism(s) by which transport of such a substrate and size diversity can be achieved, while each ABC transporter maintains selectivity for its particular substrate, represents an intriguing and yet unsolved mystery. Most interestingly, some ABC proteins can not only fulfill typical ATP-dependent substrate transport. Recent data show that ABC proteins can also act as membrane-bound proteases, ion channels, signaling molecules and even regulators of ion channels.[15]

Several ABC transporters with demonstrated peptide translocation activity have been described during the last few years. The first yeast ABC transporter with peptide transport function was Ste6, which mediates the export of the peptide mating pheromone **a**-factor.[9] A structural homolog of Ste6, the human *MDR1* gene product Mdr1 or P-glycoprotein (Pg-p), otherwise implicated in multidrug resistance development,[13] can also pump numerous hydrophobic peptides (see under II-A). Likewise, even yeast multidrug resistance ABC transporters such as Pdr5[16,17] can mediate transport of peptide substrates.[18] Additional ABC proteins that mediate peptide transport across mammalian membranes include the heterodimeric Tap1/Tap2 ABC transporter of the ER membrane (see also chapter 4). The Tap1/Tap2 ABC transporter translocates antigenic peptides, which of course lack a typical signal peptide, from the cytoplasm into the ER lumen before their cell surface presentation in association with the MHC class I molecules.[19]

Cellular viability in all eukaryotic organisms is strictly dependent on a functional organelle biogenesis. The function of organelles requires both proper targeting and assembly of nucleus-encoded, cytoplasmically synthesized proteins. Transport of proteins from their site of synthesis to their sites of action involves specific recognition by both targeting and import machineries that mediate membrane assembly and protein translocation across membranes.[20]

For instance, the signal sequence-dependent protein import machinery into mitochondria has been intensively studied. Signal sequences and numerous proteins required for protein targeting to distinct submitochondrial locations have been identified.[21,22] The matrix-targeted precursors, for instance, appear to follow a common import pathway: after their synthesis on cytoplasmic polysomes, the precursor proteins are prevented from folding into functional enzymes in the cytoplasm by binding to heat-shock proteins of the hsp70 family. Import receptors on the mitochondrial surface recognize the precursors, which are subsequently translocated to the matrix side of the inner membrane at contact sites between inner and outer membrane. The unfolded precursors bind to a mitochondrial hsp (mhsp) 70 family member, from which they are released and subsequently folded within the matrix. Finally, the positively charged N-terminal targeting signal sequence on most mitochondrial protein precursors is proteolytically cleaved during import.[21,22]

Peroxisome biogenesis and proliferation also involves direct protein import into the organelle across the single lipid bilayer.[23-25] Protein unfolding may not be necessary for import, as there is evidence that functional proteins can be imported into peroxisomes.[26,27] Known peroxisomal targeting signals include PTS1, the carboxy-terminal consensus tripeptide SKL, and PTS2, the N-terminal consensus sequence RLX$_5$HL (see also under III-C for more details). However, the molecular machinery for peroxisomal protein import and membrane assembly of peroxisomes remains to be identified.

In yeast, most of the resident vacuolar proteins reach this organelle through the secretory pathway via the Golgi network.[5,6] However, there seems to exist a

Table 2.1. The family of yeast ATP binding cassette proteins

ABC Protein	Substrate(s)	Length	Topology	Location	Phenotype	Reference
Ste6	a-factor	1290	$(TMS_6-ABC)_2$	PM, GV, ES	Sterile	10,11
Hst6 (*Candida*)	?	1323	$(TMS_6-ABC)_2$	?	?	1141746
Yor1	Oligomycin	1477	$(TMS_6-ABC)_2$	PM	Drug[HS]	153
Pmd1 (*S. pombe*)	Drugs	1362	$(TMS_6-ABC)_2$	?	Drug[HS]	165
Pdr5/Sts1/Ydr1/Lem1	Drugs, steroids	1511	$(ABC-TMS_6)_2$	PM	Drug[HS]	16, 17, 159, 160
Snq2	Drugs, steroids	1501	$(ABC-TMS_6)_2$	PM	Drug[HS]	152
Pdr10	?	1564	$(ABC-TMS_6)_2$	?	?	161
Pdr11	?	1411	$(ABC-TMS_6)_2$	PM (?)	?	162
Pdr12/LPE14c	?	1511	$(ABC-TMS_6)_2$	?	?	161
Pdr15	?	1529	$(ABC-TMS_6)_2$	?	?	161
Hba2 (*S. pombe*)	Brefeldin A	1530	$(ABC-TMS_6)_2$	?	Drug[HS]	166
Cdr1 (*Candida*)	Drugs	1501	$(ABC-TMS_6)_2$	?	Drug[HS]	167,168
Cdr2 (*Candida*)	Drugs	1499	$(ABC-TMS_6)_2$	?	Drug[HS]	169
Ycf1	GSH-conjugates/Cd	1515	$(TMS_6-R-ABC)_2$	Vac	Cd[HS]	112
Hmt1 (*S. pombe*)	Phytochelatins/Cd	830	TMS_6-ABC	Vac	Cd[HS]	117
Atm1	?	694	TMS_6-ABC	Mit	Slow-growth	124
Ssh2/Pal1/Pxa1/Pat2	LC-fatty acids	870	TMS_6-ABC	PX	Oleate-	86,147,148,151
YKL741/Pxa2/Pat1	LC-fatty acids	853	TMS_6-ABC	PX	Oleate-	149,150,151
Ssh1/Mdl2	?	820	TMS_6-ABC	?	Viable	86,126

Mdl1	?	696	TMS6-ABC	?	Ribo? Cyto?	Viable	126
Yef3	Drugs	1044	TMS2-ABC2		?	Essential	163
Adp1	?	1049	TMS2-ABC-TMS7			Viable	125
Gcn20	?	752	n.d.		Cyto?	Viable	164
YLL048c	?	1661			?	?	Z73153
YLL015w	?	1559			?	?	Z73120
YHL035c	?	1592			?	?	731612
YOR011w	?	1394			?	?	Z74919
YNR070w	?	1333			?	?	Z71685
YKR103w	?	1218			?	?	549649
YPL226w	?	1196			?	?	Z73582
YOL075c	?	1095			?	?	Z74817
YNL014w	?	1044			?	?	1301836
YER036p	?	610			?	?	603269
YDR091c	?	608			?	?	914875
YDR061w	?	539			?	?	798912
YKR104w	?	306			?	?	Z28329
YFL028c	?	289			?	?	836726

In cases where no published references are available, the EMBL/Genbank sequence accession or the NCBI identification numbers are listed to allow for sequence retrieval. Abbreviations are: *PM*, plasma membrane; *Vac*, vacuole; *Cyto*, Cytoplasm; *GV*, Golgi vesicle; *ES*, endosome; *Ribo*, ribosome; *PX*, peroxisome; *Mit*, mitochondria; *R*, regulatory domain; *TMS*, transmembrane segment; *ABC*, ATP binding cassette; *HS*, hypersensitivity; *LC*, long chain; *GSH*, glutathione; *n.d.*, not determined direct protein translocation mechanism from the cytoplasm to the vacuole operating independently of the classical secretory pathway.[28,29] For instance, the direct cytoplasm to vacuole transport of the vacuolar aminopeptidase I requires an N-terminal amphipatic α-helix as a signal sequence which does not resemble a typical signal peptide for ER-translocation.[30,31] Another vacuolar membrane protein, α-mannosidase, also assembles at the luminal face of the vacuolar membrane via specific or dedicated transport system(s), none of which has been identified to date.

The scope of this chapter is to summarize recent developments in unusual protein secretion and intracellular targeting in yeast. In particular, we will recapitulate the mechanism of a-factor pheromone export, which is mediated by the Ste6 ABC transporter, and operates independently of the classical secretory pathway. Furthermore, we shall discuss additional novel protein export pathways that are distinct from both the secretory pathway *and* independent of Ste6. Moreover, we will also discuss instances where organelle biogenesis involves protein trafficking and membrane translocation events that are different from the known mechanisms for protein targeting such as the vesicular exocytic and endocytic pathways. Recent discoveries in direct cytoplasm to organelle protein import and the discovery of numerous ABC transporters in various distinct cellular membranes (Fig. 2.2)[59] prompted us to review facts and myths about the role of ABC transporters in intra- and intercellular protein transport routes that do not use conventional cleavable signal peptide mediated transport. We will discuss a possible role of a newly emerging subfamily of yeast ABC transporters in transmembrane peptide transport, which appears to be located in certain intracellular organelle membranes (Fig. 2.2).

II. UNUSUAL PROTEIN EXPORT ACROSS THE YEAST PLASMA MEMBRANE

A. Yeast a-Factor Pheromone Export Requires a Dedicated ABC Transporter

The yeast *Saccharomyces cerevisiae* is a unicellular eukaryotic organism, which exists in three distinct cell types. Conjugation or mating of two haploid cells, *MATa* and *MATα*, results in the formation of a diploid a/α cell.[32,33] Mating requires the action of extracellular peptide hormones known as the mating pheromones a-factor and α-factor, that are released from haploid *MATa* and *MATα* cells, respectively. The pheromone produced by *MATα* cells, α-factor, is a 13-residue peptide which is proteolytically liberated from a larger glycosylated precursor, and secreted via the classical ER-Golgi secretory pathway.[33]

Mating pheromone a-factor is synthesized as a 36-residue and 38-residue precursor from the *MFa1* and *MFa2* genes, respectively.[34] The precursors are

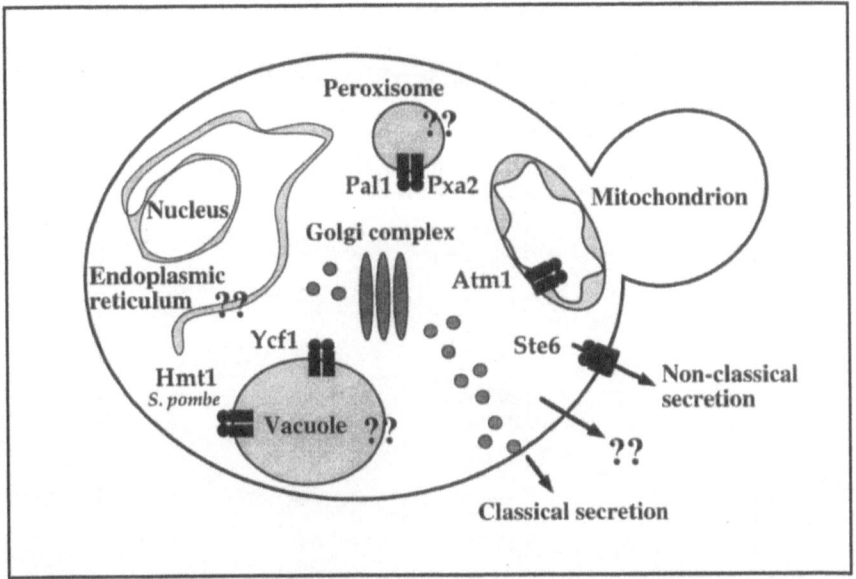

Fig. 2.2. Yeast ABC transporters involved in unusual secretion and organelle biosynthesis and/or possibly in peptide translocation. Evidence for nonvesicular peptide secretion or translocation (??); for further details see text.

proteolytically processed and posttranslationally modified in the cytoplasm.[35-37] Prior to its release to the extracellular space, a-factor is further converted to a mature, bioactive 12-residue lipopeptide, whose C-terminal cysteine residue is both farnesylated and carboxymethylated.[38] The signal triggering these modifications was identified as the CAAX box, where C is cysteine, A is an aliphatic residue and X may be any amino acid, found at the extreme C-terminus of all prenylated proteins, including both yeast and mammalian Ras proteins[39] and the unmodified a-factor precursors.[40]

The players in a-factor biogenesis and the export model

The biogenesis of active extracellular a-factor requires several gene products. First, two structural genes, *MFa1* and *MFa2*, encode the known pheromone precursors.[34] Furthermore, a farnesyl transferase activity, several processing proteases, a carboxymethyltransferase and a dedicated pheromone transporter are required for a-factor biogenesis. A hypothetical and somewhat speculative working model for a-factor pheromone secretion by haploid *MATa* cells is depicted in Figure 2.3. Early events such as membrane association and posttranslational processing may precede export of mature a-factor which occurs at the plasma membrane.

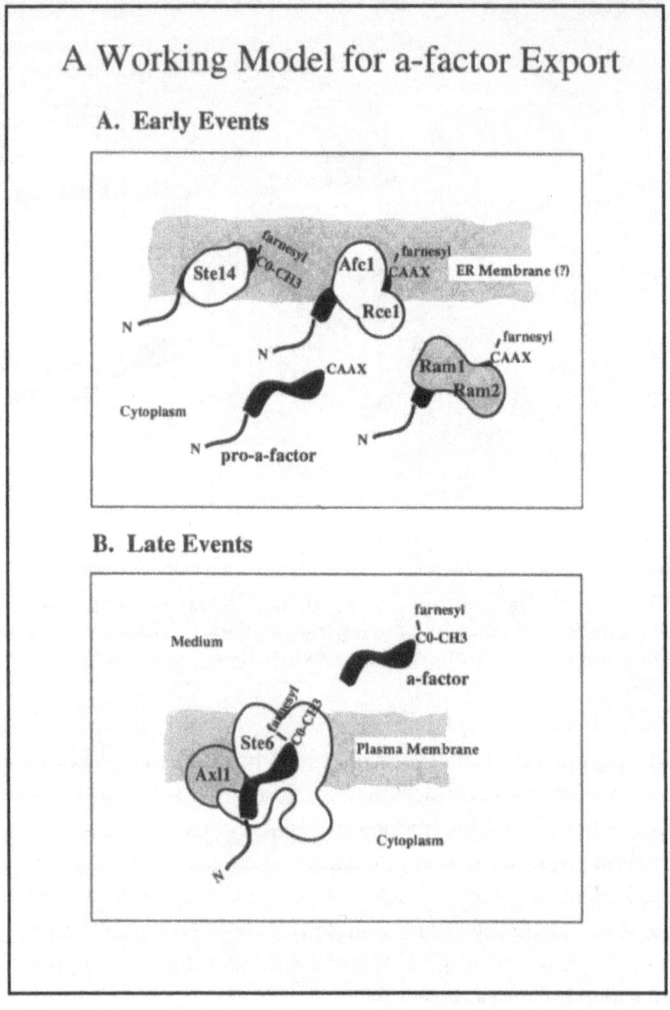

Fig. 2.3. A hypothetical working model for yeast mating pheromone a-factor export. The Ste6-dependent **a**-factor secretion apparatus may be functional in the plasma membrane as a transport-competent translocation complex. *(A) Early events.* Pro-**a**-factor is synthesized in the cytoplasm as an inactive precursor that is farnesylated at the C-terminal CAAX box by a heterodimeric farnesyltransferase (Ram1, Ram2). Farnesylation targets the precursor to a membrane where it undergoes proteolytic removal of the three C-terminal residues of the CAAX box (Afc1, Rce1; J. Rine; personal communication). Modification of the CAAX box is completed after carboxymethylation of the C-terminal cysteine by a methyltransferase (Ste14). *(B) Late events.* It has not been established when exactly the processing protease Axl1 cleaves the pro-peptide at the **a**-factor N-terminus. However, it may precede the actual translocation step, which is mediated by the Ste6 ABC transporter in the plasma membrane.

The proteolytic processing activities required for **a**-factor maturation have been characterized at the biochemical level[36,37] and the corresponding genes have been recently identified. The *AXL1* gene product, which also plays a role in bud site selection,[41] was identified as one of the pro-**a**-factor processing proteases.[35] In addition, two novel genes, *RCE1* and *AFC1*, required for efficient proteolytic CAAX box processing, which must precede carboxymethylation, were recently isolated (J. Rine; personal communication).

Initially, **a**-factor is synthesized on cytoplasmic polysomes as an unglycosylated precursor that lacks a hydrophobic signal peptide but has a hydrophilic N-terminal extension.[11,34] After its biosynthesis, pro-**a**-factor is lipid-modified at the CAAX box by the heterodimeric Ram1/Ram2 farnesyltransferase.[42,43] CAAX box prenylation most likely targets the pheromone to an intracellular membrane for further maturation. The three C-terminal amino acids of the CAAX box are then clipped through *RCE1* and *AFC1*-mediated proteolysis (J. Rine; personal communication), before the pheromone can become carboxymethylated by Ste14.[44-46] Hydrophobicity analysis of the *STE14* gene,[47] encoding the functional **a**-factor carboxymethyltransferase,[44] suggested that Ste14 is membrane associated.

Interestingly, Afc1 is a membrane protein which may be localized to the ER membrane, because it has a KKXX consensus ER retrieval signal[48-50] at the extreme C-terminus (J. Rine; personal communication) This raises the interesting possibility that **a**-factor precursor processing could take place, at least to a certain extent, on the cytoplasmic face of the ER membrane (J. Rine; personal communication). Limited protease digestion experiments confirm that newly synthesized pheromone does not enter vesicular compartments and remains exposed to the cytoplasm, since the pheromone is highly susceptible to protease treatment after its synthesis.[51] Preliminary results from immunolocalization experiments also suggest a ER localization for the Ste14 methyltransferase, although Ste14 does not carry a consensus KKXX ER retrieval signal (J. Rine; personal communication).

The actual translocation of the **a**-factor pheromone, however, is most likely to occur at the plasma membrane, where Ste6 mediates pheromone extrusion directly from the cytoplasm or from the cytoplasmic aspect of the plasma membrane. It is unclear where Axl1, which appears to be a membrane protein of unknown subcellular location,[35] is functional. Its role in N-terminal **a**-factor processing, however, could indicate involvement in late events in pheromone secretion (Fig. 2.3). Thus, the mechanism of pheromone export may involve a "piggy-back" mechanism, in which **a**-factor travels to the cell surface, starting at the cytoplasmic face of the ER membrane where initial processing occurs, before Ste6-mediated transport across the plasma membrane. The Ste6 pheromone ABC transporter was indeed functionally localized to the plasma membrane.[52] Most interestingly, however, the bulk of Ste6 appears to be present in a intracellular Golgi-like vesicular compartment as determined by immunofluorescence experiments[52] and biochemical fractionation studies.[53] In addition, proteolytic turnover measurements proved Ste6 an extremely short-lived protein and showed that its plasma membrane localization is only transient.[53,54]

Compelling genetic and biochemical evidence suggested that bioactive a-factor is translocated from the cytoplasm across the plasma membrane[52] to the extracellular space in a rate-limiting export step mediated by Ste6.[11] Export of a-factor apparently bypasses the classical ER-Golgi secretory pathway, because extracellular pheromone is still being produced when *MATa* cells carrying temperature-sensitive secretion-defective (*sec*) mutations are shifted to the restrictive temperature.[10,11,51] Thus, a-factor secretion apparently does not require a vesicular secretory compartment and therefore represents a novel route for protein secretion in eukaryotic cells.[8,9] The order of events in a-factor maturation was also confirmed in vitro by using a-factor maturation assays.[55] It is unclear at present why only fully processed a-factor can be efficiently secreted from cells,[11,44,56] but one may assume that a-factor release is coupled to its intracellular processing, because very little, if any, mature a-factor can be detected intracellularly.[11,44]

However, it is very important to point out that the Ste6 pheromone transporter itself is targeted to the cell surface via the ER-Golgi secretory pathway.[53,54] How then can one explain the paradox observation that blocking the secretory pathway at all different stages[51] does not impair a-factor export? The *sec*-independence could be explained if there was a large cellular pool of a-factor pheromone, maybe in the plasma membrane, which cannot be depleted during the temperature shift to block exocytosis. Another more likely possibility would be that a significant amount of extracellular a-factor pheromone stays cell-associated for a period of time that would persist the time of blocking of the secretory pathway. Thus, even after stopping vesicular fusion at the cell surface, there would be sufficient pheromone at the cell surface able to confer mating competence. This idea is supported by the extreme insolubility of a-factor in aqueous solutions and by the low diffusibility of the pheromone.[51]

Recognition of pheromone by Ste6

One may assume that pheromone interaction with Ste6 would presumably take place at the inner aspect of the plasma membrane. This would require distinct structural features such as a-factor binding sites on the cytoplasmic loops in Ste6. Indeed, a significant homology of the Ste6 linker region between the first ABC domain and TMS7 to the third extracellular loop of Ste3 a-factor receptor suggested this to be a potential pheromone binding site.[11] However, partial deletion of this linker region does not destroy a-factor transport function of Ste6 (R. Kölling, personal communication).

More recently, site-directed mutagenesis within the first ABC domain of Ste6 revealed the importance of this cytoplasmic loop in substrate recognition and transport.[57] Missense mutations introduced near or especially within the conserved LSGGQ motif of the first Ste6 nucleotide-binding domain caused severe a-factor transport defects. Importantly, the authors could demonstrate that the mutations do not alter steady-state level and/or subcellular localization of Ste6.[57] Another possible explanation is that mutations in the highly conserved domains influence the overall structure of Ste6 in the membrane, which in turn could affect pheromone transport efficiency and/or pheromone binding.

The binding of the a-factor pheromone to Ste6 could also take place within the phospholipid bilayer, since farnesylated a-factor is extremely hydrophobic and associates with phospholipid bilayers even in the absence of Ste6 (K. Kuchler, unpublished results). Moreover, the farnesyl moiety of a-factor has been shown to promote association with artificial phospholipid bilayers in vitro,[58] and it can mediate membrane association of a chimeric interleukin-1α:a-factor fusion protein.[59] Hence, lateral diffusion of pheromone in the membrane could bring a-factor to the Ste6 transporter before binding and extrusion to the medium. Notably, variant a-factor genes were produced to generate isoprenylation targets for geranyl-geranylation, CVIL and SVCC, replacing the normal CVIA farnesylation motif.[60] Geranyl-geranyl-modified a-factor variants retain their bioactivity and are exported by a Ste6-dependent mechanism. Thus, although not farnesyl-specific, membrane association of a-factor and its recognition by Ste6 clearly requires a lipid modification such as CAAX box prenylation.

A somewhat similar model for substrate binding and extrusion, the "molecular vacuum cleaner model," was suggested for the mammalian P-glycoproteins (Pg-p) drug efflux pumps.[13,14] In this model, it was proposed that hydrophobic drugs, in contrast to hydrophilic drugs, actually never enter the cell, since these drugs would partition into the phospholipid bilayer due to their hydrophobicity. Lateral membrane diffusion and interaction of drugs with certain transmembrane α-helices of Pg-p would precede ATP-dependent drug extrusion. Strikingly, numerous mutations in predicted membrane-spanning domains of Pg-p lead to nonfunctional transporters or to a different substrate specificity,[14,61] supporting the vacuum cleaner model. In the case of a-factor, one may also speculate that accessory proteins such as plasma membrane bound receptors, which would bind prenylated a-factor, could facilitate the interaction of a-factor with its export machinery and/or membrane association of the maturing pheromone. Indeed, there is preliminary evidence that farnesylated IL-1α:a-factor chimeras are found exclusively in membrane fractions including the plasma membrane.[59] Similarly, a heterologous protein A:CAAX box fusion protein is exclusively targeted to the plasma membrane in mammalian MDCK cells, but not to other cellular membranes, suggesting the existence of plasma membrane receptors for prenylated proteins.[62] However, high affinity binding sites for prenylated proteins have only been identified in rat liver microsomal membranes.[63]

What is the secretion signal for pheromone export?

The actual secretion signal that drives a-factor export is poorly understood. It has been speculated that the hydrophilic N-terminal extension of pro-a-factor promotes pheromone release to the medium, but the existence of many mutations within the N-terminal leader sequence with no obvious phenotype seems to contradict this idea.[64] In addition, expression of a variant a-factor gene lacking this N-terminal extension, still produces extracellular pheromone (J. Rine; personal communication).

The farnesylated and carboxymethylated C-terminus of a-factor represents another potential secretion signal. Our own studies with murine interleukin-1α:a-factor fusion proteins expressed in yeast seem to support this view.[59] Mature murine interleukin-1α (IL-1α) is involved in a variety of immunoregulatory responses.[65,66] Like yeast a-factor, mammalian IL-1α lacks a hydrophobic signal peptide (see also chapter 3), and it is secreted from mammalian cells bypassing the classical secretory pathway.[67] The cytokine remains cytoplasmic when expressed in yeast, unless a hydrophobic signal peptide is attached to its N-terminus leading to its secretion.[68] By contrast, a chimeric IL-1α:a-factor pheromone expressed in a *MATa Δmfa1 Δmfa2* yeast strain is quantitatively associated with membranes, presumably because C-terminal CAAX box is farnesylated in vivo and therefore mediates membrane insertion.[59] A *MATa Δmfa1 Δmfa2* cell, which lacks both structural genes for a-factor, is absolutely sterile, due to the lack of a-factor pheromone that is essential for mating.[69] Surprisingly, quantitative mating assays demonstrated that the chimeric IL-1α:a-factor pheromone has retained at least some a-factor bioactivity, since it is able to mediate conjugation.[59] Thus, the chimeric pheromone must be at the extracellular space or on the outside of the cell. Although the extremely sensitive mating assay allows for detection of very low levels of extracellular pheromone, cell lysis cannot explain the appearance of extracellular IL-1α:a-factor chimera. IL-1α:a-factor-mediated conjugation is solely dependent on a functional Ste6 transporter, as it does not occur in the isogenic *Δste6* mutant strain.[59] Another possible explanation for the appearance of pheromone in the medium could be intracellular cleavage of the IL-1α:a-factor chimera that would generate aberrant intracellular a-factor forms recognized and transported by Ste6. This seems unlikely, however, because at least one chimeric pheromone construct is stable and does not give rise to unspecific degradation products.

These results demonstrate that the size requirements for pheromone export by Ste6 may be more relaxed than previously anticipated, since Ste6 is apparently able to transport even an enlarged "a-factor pheromone" such as the 17 kDa chimeric IL-1α:a-factor protein (compared to the 3 kDa of mature a-factor).[59] These results suggest that the fully lipid-modified C-terminal CAAX box in a-factor may represent a pheromone secretion signal, since it can confer targeting to the extracellular space in a Ste6-dependent manner. Indeed, a C-terminal CAAX box can target bacterial protein A to the plasma membrane in mammalian MDCK cells.[62] Moreover, farnesylation, but also carboxy-methylation, and a polybasic domain near the C-terminus seems essential for plasma membrane targeting of Ras proteins.[62,70]

The farnesyl moiety of a-factor has also been speculated to mediate pro-tein-protein and protein-lipid interactions.[58] Thus, the hydrophobic farnesyl tail of the pheromone at the cell surface could facilitate initial contact of mating cells. The farnesyl moiety of a-factor is essential for mating pheromone bioactivity[71] and the pheromone potency may be determined by the lipid modi-fication.[72] An in vivo mating restoration assay demonstrated that synthetic a-factor,[73] when exogenously added to mating cells, can restore, albeit very in-efficiently, conjugation. Mating is only observed if a-factor is added to cell mix-tures containing *MAT*a *Δmfa1 Δmfa2* and *MAT*α cells, but not when added to *MAT*a *Δmfa1 Δmfa2 Δste6* and *MAT*α cells.[71] This finding indicates that Ste6, in addition to its essential role in a-factor pheromone export, could have further yet unknown functions in mating. One could also speculate that Ste6 actually translocates a-factor to the cell surface without efficiently releasing the phero-mone from the cell surface to the medium. This way, Ste6 would somehow present cell surface-associated a-factor to mating-competent *MAT*α cells. Re-markably, mutations in *Ste6* were recently identified which affect zygote for-mation but not pheromone export per se, confirming a possible novel function for Ste6 in cell fusion.[74]

Membrane topology and energetics of Ste6

The hypothetical model for the a-factor export assumes that the predicted membrane topology of Ste6, which places N- and C-terminus and both ABC domains at the cytoplasmic face of the plasma membrane, is correct.[11] Such a Ste6 topology was indeed demonstrated for the N-terminal half of Ste6. Fusion of topogenic reporter genes such as alkaline phosphatatese and invertase to N-terminal Ste6 TMS were expressed in *E. coli* and in *S. cerevisiae*, respectively.[75] Accessibility of the reporter genes to posttranslational modifications such as glycosylation of invertase and the enzymatic activity of alkaline phosphatase in the extracellular space was consistent with the predicted six TMS organization of the N-terminal half of Ste6.[75] Limited protease digestion experiments of func-tional Ste6 derivatives that carry specific Factor Xa protease cleavage sites in both intra- and extracellular hydrophilic loops within the first half of Ste6 further support the predicted $(ABC-TMS_6)_2$ topology of Ste6 for at least two trans-membrane spans (V. Huter and K. Kuchler, unpublished results). Moreover, linker insertion and point mutations in predicted intracellular hydrophilic loops of Ste6 lead to loss of function, although it is not known whether the same mutations also cause a mislocalization of Ste6 (V. Huter and K. Kuchler, unpublished results).

From the energetic point of view, it seems clear that the **a**-factor transloca-tion step across the plasma membrane requires energy. Indeed, Ste6 has been shown to be an ATP-binding protein in vitro,[52] and mutations in the ABC do-mains that impair ATP-binding also debilitate Ste6 function,[56,57] suggesting that ATP-hydrolysis powers pheromone export. In addition, both ABC domains are required for Ste6 function, because only both halves of Ste6 co-expressed in the same cell, but not the N-terminal TMS_6-ABC or C-terminal $ABC-TMS_6$ half alone, give rise to a functional pheromone transporter.[56] It is not known, how-ever, if ATP-binding and/or ATP-hydrolysis also serves a structural function for the architecture of a transport complex in the membrane, as was previously shown for the certain bacterial ABC transporters.[76] Taken together, both Ste6 halves and their ABC domains must tightly interact to form a putative mem-brane pore through which pheromone extrusion can occur. This interaction is quite similar to the assembly of the mammalian Tap1 and Tap2 ABC transport-ers of antigen presentation. Their interaction results in a Tap1/Tap2 heterodimer in the ER membrane, and leads to an antigen transporter that translocates 8-12-residue peptides into the ER lumen (see also chapter 4). The physical in-teraction of Tap1 and Tap2 must be very strong, since it can only be disrupted by sodium dodecylsulfate treatment, and because both proteins always co-purify as a complex. In addition, the heterodimeric Tap1/Tap2 complex can be immu-noprecipitated using monoclonal antibodies recognizing either Tap1 or Tap2.[77]

Regulation of expression and the intracellular trafficking of Ste6

The expression of the Ste6 transporter is under hormonal control, as α-factor-arrested cells express almost 10-times higher levels of Ste6.[52] In vegeta-tively growing cells, Ste6 is present predominantly in intracellular vesicles that subtend the plasma membrane.[52,53] Exposure of *MATa* cells to α-factor, how-ever, induces Ste6 expression and leads to an apparent redistribution of Ste6 from internal vesicular pools to the cell surface.[52] Simultaneously, Ste6 becomes concentrated in the tip of the mating projection.[52] This morphological change, also known as "schmoo," is always directed toward the mating partner and in-duced by pheromones.[32,78] Schmoo-formation indicates a functional mating signal transduction cascade and ensures that mating partners are arrested in the G1 phase of their cell cycles.[32] At the same time, the pheromone signal in-duces a number of biochemical changes considered a prerequisite for fusion of haploid cells, which occurs exclusively at the schmoo-tips.[33,78] The polarized Ste6 localization must result in a pronounced anisotropy of a-factor secretion around the tip of the projection.[52] In addition, the hydrophobicity of a-factor leads to a steep pheromone concentration gradient reaching from the cell surface into the medium. Such a delocalized pheromone signal, and the highest possible levels of extracellular pheromone (as caused by increased Ste6 expression), is believed to be necessary for courtship during mating partner discrimination.[79]

The Ste6 transporter is a short-lived protein with an apparent half-life of only 10-15 minutes.[53,54] In addition, the metabolic stability of Ste6 is dramatically increased in a *pep4* vacuolar protease mutant strain,[53,54] suggesting that endocytosis delivers Ste6 from the cell surface to the vacuole. Indeed, vacuolar degradation of Ste6 is dependent on a functional endocytic pathway.[53,54] Likewise, constitutive and pheromone-induced endocytosis has been shown to be responsible for uptake and vacuolar delivery of the pheromone receptors Ste3 (a-factor receptor) and Ste2 (α-factor receptor) via endosomes.[80,81] Notably, endocytosis of both Ste2[82] and Ste3[83] seems triggered by their ubiquitination at the cell surface. Similarly, Ste6 is ubiquitinated in vivo[53] and Ste6 accumulates in an ubiquitinated form in the plasma membrane of *end4* mutant cells[53] blocked in the first step of endocytosis, the formation of early endosomes.[84]

The rapid turnover of Ste6 also provides a plausible explanation for the polar localization of Ste6 in the plasma membrane of pheromone responsive cells.[52] While all of Ste6 present in cells before α-factor exposure is degraded with a half-life of 10-15 min,[53] newly synthesized, pheromone-induced Ste6 molecules are deposited mainly at the site of membrane growth along the mating projection tip.[52] This way, a predominant schmoo-tip localization of Ste6 can be achieved, while at the same time generating an anisotropy of a-factor secretion.[52] In other words, a 10-fold induced Ste6 expression and its deposition at the schmoo-tip in response to pheromone, can easily compensate for the endocytic removal of Ste6 from the cell surface outside the mating projection. Due to kinetic arguments, it seems unlikely that the polar Ste6 localization is caused by increased Ste6 degradation or increased endocytosis at cell surface areas outside the schmoo-tip.[54] Moreover, there is preliminary evidence that Ste6 turnover does not change in α-factor pheromone-induced cells when compared to normal cells (R. Kölling, personal communication). However, the intracellular vesicular staining of Ste6 appears to decrease in pheromone-arrested cells,[52] implying enhanced surface delivery of Ste6-containing vesicles upon pheromone induction.[52]

The Ste6 presence in intracellular vesicles in normal cells was previously interpreted to be due to exocytic secretory organelles such as Golgi or secretory vesicles that contain Ste6 en route to the plasma membrane.[52] This idea is in agreement with a report demonstrating a Golgi localization for Ste6.[53] Likewise,

Ste2, the α-factor receptor of *MAT*a cells, shows an identical localization pattern as Ste6, both in normal and in pheromone-treated cells.[85] However, because of the endocytic Ste6 trafficking to the vacuole, the vesicular spot and dot-like staining of Ste6 in uninduced yeast cells[52,53] could also be due, at least in part, to the presence of Ste6 in endocytic compartments.

Although it is established that Ste6 mediates the actual **a**-factor transport step across the plasma membrane, it cannot be ruled out that additional yet unknown proteins perhaps dynamically associate with Ste6 to build up a **a**-factor translocation complex. To be able to more precisely define the **a**-factor binding domains in Ste6, and to define the function of other proteins required for **a**-factor biogenesis, including Axl1, Ste14 and Ram1/Ram2, Afc1 and Rce1, more Ste6 transporter and processing mutants will have to be isolated and characterized in the future. These studies will be greatly facilitated by the availability of a reconstituted in vitro **a**-factor transport system. Studies along this line will help to answer many remaining questions concerning the molecular mechanism of Ste6-mediated of **a**-factor export.

Ste6 is a typical member of the ABC transporter family being most closely related in primary sequence and predicted membrane topology to the mammalian P-glycoproteins.[10,11] In fact, both human Mdr1[86] and mouse Mdr3[87] when functionally expressed in yeast can partially compensate for the loss of Ste6 function. Likewise, MRP (multidrug-resistant-related protein)[88] and the *Plasmodium falciparum* multidrug ABC transporter Pfmdr1[89] are able to partially restore mating in sterile *ste6* null mutants, respectively.[89,90] These data demonstrate the intrinsic potential of eukaryotic ABC transporters for peptide translocation in vivo. Moreover, overexpression of P-glycoprotein confers resistance to certain peptide antibiotics such as dolastatin,[91] protease inhibitors[92] and ionophores like valinomycin.[93] Interaction of several peptides with the Mdr1 transporter takes place at binding site(s) distinct from those for verapamil and azidopine.[94] It was, therefore, suggested that export of hydrophobic peptides may be a possible physiological function of mammalian P-glycoproteins, although no direct proof for this idea is available until now.[61,86] However, it should be emphasized that the physiological function(s) of mammalian P-glycoproteins are still far from being firmly established, despite accumulating evidence for a function in intracellular phospholipid[95-97] and/or steroid transport[98] in vivo.

B. Unusual Protein Export Independent of the Ste6 ABC Transporter

The yeast Ste6 pheromone pump was the first eukaryotic ABC pump whose physiological substrate, the **a**-factor pheromone, could be identified.[8] Until recently, this was the only known nonclassical protein export pathway existing in *Saccharomyces cerevisiae*. In mammalian cells, an increasing number of secreted proteins with physiological importance are identified that lack a typical secretory signal sequence.[8,86,99] For example, the cytokine interleukin-1β,[67] thioredoxin,[100] FGF-1 (*fibroblast growth factor*) and FGF-2[101] or galectin-1[102] are actively secreted despite the lack of a typical signal peptide. Mammalian galectin-1, a 14 kDa homodimeric lactose-binding lectin that lacks posttranslational modifications other than an acetylated N-terminus, was recently also expressed in yeast.[103] Intriguingly, mammalian galectin-1 is exported via the yeast plasma membrane bypassing the secretory pathway.[103] Furthermore, secretion of galectin-1 from yeast does not require the Ste6 a-factor pheromone transporter for its export.

A genetic screen to identify yeast genes for galectin export allowed for the cloning of *nce1* and *nce2* (for *nonclassical export*).[103] While *NCE1* encodes a small protein of 53 amino acids of yet unknown function, Nce2 is a 173-residue protein with a predicted molecular mass of 19 kDa and four potential TMS. Nce2 has no significant homology to any known protein in databases. Deletion of the *nce2* gene is not detrimental to yeast viability, although it leads to a drastic reduction of galectin-1 export.[103] This confirms the link between Nce2 and another novel Ste6-independent non-classical protein secretion pathway in *Saccharomyces cerevisiae*.

Furthermore, an additional novel hydrophilic protein of 25 kDa named Nce3 was identified.[103] Nce3 appears to represent an endogenous substrate for nonclassical secretion in yeast. Nce3 export is independent of the secretory pathway, as shown with a mutant defective in the *sec14* gene.[103] Notably, Nce3 secretion also does not require the function of either Ste6 or Nce2. These results suggest that yeast cells contain additional nonclassical protein export machineries in addition to the Ste6 and the Nce pathways. Further work will be necessary to uncover the obvious role of Nce2 in nonclassical protein export. Nce2 may be a subunit of a multimeric export complex in the plasma membrane, as suggested by the presence of several predicted membrane-spanning domains in Nce2.[103] Alternatively, one can speculate that Nce2 is an accessory protein for a core component of the non-classical export machinery or the role of Nce2 may just be indirect. It will be of great interest to identify the gene products possibly interacting with Nce2 and required for Nce3 secretion to uncover this novel but still unknown protein export mechanism. Moreover, these yeast components and/or genes may become important tools for the identification of functional analogs of non-classical export pathways from mammalian cells.

III. UNUSUAL PROTEIN TRANSLOCATION ACROSS INTRACELLULAR MEMBRANES

The biogenesis of membrane-surrounded subcellular compartments in all eukaryotic cells requires the import of cytoplasmically synthesized proteins into the organelle lumen as well as the assembly of organellar membranes. Protein import can occur via vesicle fusion or by direct translocation across the organelle membranes. In the case of direct protein import, transport of proteins to their sites of action involves the specific recognition by an import machinery. Several translocated proteins from different cellular compartments and even their targeting signals were uncovered during the recent years. However, except for the mitochondrial protein import, the translocation mechanisms remain largely unknown. In the following section, we will summarize examples of direct protein translocation into yeast organelles and discuss a possible role for ABC transporters in this process.

A. Vacuole

The yeast vacuole is equivalent of the mammalian lysosome, in which the bulk of cellular protein turnover is accomplished. Proteins destined for degradation can enter the vacuole via endocytosis[104,105] and autophagocytosis.[106,107] The majority of vacuolar proteins are transported to the vacuole through the classical secretory pathway.[5,6] However, certain resident vacuolar proteins such as α-mannosidase and aminopeptidase I are imported into the vacuole directly from the cytoplasm bypassing ER-to Golgi transport through the secretory pathway.[28,29]

The vacuolar α-mannosidase is a multimeric enzyme comprised of a 107, 73 and a 31 kDa polypeptide subunit. All three polypeptides are derived from processing of a single gene called *ams1*. The 73 and 31 kDa subunits are generated from the 107 kDa species by proteolytic cleavage within the vacuole. Although α-mannosidase is not an integral membrane protein, it is peripherally associated with the luminal face of the vacuolar membrane.[28] The protein is not glycosylated despite the presence of seven potential N-glycosylation sites in the *ams1* gene, and there is no apparent signal peptide for ER-translocation. Newly synthesized α-mannosidase is delivered to the vacuole even when the secretory pathway is blocked by a subset of *sec* mutations.[28]

A detailed characterization of α-mannosidase biosynthesis suggested that the protein is sequestered directly from the cytoplasm into the vacuole but not via the secretory pathway. However, attachment of a cleavable carboxypeptidase Y (CPY) signal sequence to the N-terminus of α-mannosidase leads to the appearance of a glycosylated protein, suggesting that only the variant CPY:Ams1, but not the normal Ams1 enzyme, can enter the ER lumen.[28] Further, experiments with C-terminal deletions of α-mannosidase fused to invertase suggested that the targeting sequence for Ams1 import into the vacuole is located within the last 157 amino acids of its C-terminus. Overexpression of native α-mannosidase revealed that the import machinery for α-mannosidase is saturable.[28] While these results clearly suggest a mechanism for α-mannosidase import into the vacuole independent of the secretory pathway, individual components required for Ams1 membrane translocation have not been identified.

Yeast aminopeptidase I is a 600 kDa soluble multimeric leucine aminopeptidase of the vacuole. The 514 amino acid precursor of aminopeptidase I, encoded by the *ape1* gene, contains a N-terminal pro-peptide, which is removed upon entry into the vacuole by vacuolar endoproteinases such as proteinase A and proteinase B.[29,30] The pro-peptide of aminopeptidase I does not resemble or act as an ER signal sequence. Like α-mannosidase, vacuolar localization of aminopeptidase I occurs independent of the secretory pathway. Maturation of the enzyme is not blocked in *sec12, sec23, sec18* and *sec7* mutants that are defective in transport through the early secretory pathway.[29] In contrast to α-mannosidase, the vacuolar targeting sequence for Ape1, which is necessary and sufficient for vacuole translocation, is found within its pro-peptide.[29,30]

Studies on the processing of pro-aminopeptidase I have identified a intermediate form of aminopeptidase I in the vacuole, which was cleaved near amino acid 17 of the pro-peptide, suggesting a signal sequence-like function for the released peptide.[30] Indeed, after expression of a mutant aminopeptidase I lacking the first 16 amino acids (Δ1-16), the protein was no longer converted to the mature form and remained in the cytoplasm. The same was observed for a mutant aminopeptidase totally lacking (Δ1-45) the pro-peptide.[30] There is evidence that the translocation of aminopeptidase I starts with the insertion of the pro-peptide into the vacuolar membrane.[30] Similar to α-mannosidase, overexpression of aminopeptidase I results in the accumulation of the precursor in the cytoplasm, suggesting the existence of a saturable import machinery in the vacuolar membrane.[29]

The predicted secondary structure of the Ape1 pro-peptide comprises a amphipatic α-helix followed by a β-turn and another α-helix, all together forming a helix-turn-helix structure.[31,108] Analysis of randomly generated mutations in the pro-peptide of Ape1 revealed the importance of the first amphipatic α-helix for proper vacuole import.[31] All mutations that affected Ape1 processing are in the predicted amphipatic α-helix in a region spanning amino acids 6-15, except a proline to leucine substitution at position 22 in the pro-peptide β-turn. Amino acid changes in the region of amino acids 6-15, which alter the hydrophobic nature of the first amphipatic α-helix, result in a complete Ape1 processing block and lead to the accumulation of mutant pro-aminopeptidase I in the cytoplasm.[31]

The function of the aminopeptidase I pro-peptide as a sorting and/or targeting signal could be brought about by a direct interaction of the amphipatic α-helix with (i) a cytoplasmic factor, (ii) the vacuolar membrane or (iii) a membrane bound receptor/protein or pump. Mutations in the amphipatic α-helix prevent binding of aminopeptidase I to the vacuolar membrane and maintain the enzyme soluble. Notably, the proline to leucine mutation at position 22 in the β-turn of the pro-peptide blocks aminopeptidase I processing but not Ape1 precursor binding to the vacuolar membrane.[31] This suggests that the Ape1 pro-peptide has several functions, including the binding of the precursor to the vacuolar membrane and mediating or facilitating subsequent step(s) of transmembrane translocation into the vacuole.

The cytoplasm to vacuole targeting of aminopeptidase I was reconstituted in vitro in a permeabilized cell system.[109] The import process is both temperature and ATP-dependent, and import requires a functional vacuolar ATPase.[109] A genetic screen yielded a collection of yeast mutants defective in the cytoplasm to vacuole protein targeting pathway of aminopeptidase I.[110] Five complementation groups were specifically defective for the targeting and translocation of aminopeptidase I. These *cvt* mutants (for *c*ytoplasm to *v*acuole *t*argeting) specifically accumulate the unprocessed Ape1 precursor in the cytoplasm, without disturbing secretion or vacuolar biogenesis via the secretory pathway.[110] Interestingly, a phenotypic and genetic overlap between autophagocytosis and cytoplasm to vacuole targeting in yeast was recently demonstrated.[111] The results suggest that import of aminopeptidase I into the vacuole shares a number of components required for bulk autophagocytosis. Aminopeptidase I specificity and its constitutive import appears to rely on additional interacting proteins. Alternatively, the aminopeptidase I maturation defect in *aut* mutants defective in autophagocytosis[107] could be a secondary effect due to alterations in the vacuole membrane composition. The future analysis of the *cvt* and *aut* gene products will hopefully provide information about the components involved in the direct cytoplasm to vacuole translocation of resident vacuolar proteins.

The Ape1 precursor was not detected inside a membrane-surrounded compartment before its vacuolar appearance, neither in the wild-type situation[29] nor in *cvt* mutants.[110] These observations make a translocation mechanism via vesicle formation and fusion from the cytoplasm as seen during autophagocytosis[106,107] or endocytosis[105] unlikely. Instead, a direct Ape1 membrane translocation mediated by a translocation complex or even a membrane pump such as an ABC transporter[59] of the vacuolar membrane appears more likely.

A vacuolar member of the yeast family of ABC transporters is the 1515-residue *ycf1* gene product (for *y*east *c*admium *f*actor), a homolog of the human cystic fibrosis transmembrane conductance regulator, CFTR.[112] Ycf1 was cloned in a multi-copy based genetic screen in which yeast cells were selected based on their ability to grow in the presence of increased cadmium concentrations.[112] One can envisage, that cadmium detoxification presumably involves vacuolar sequestration of intracellular metal ions into the vacuole. Moreover, recent data indicate that Ycf1 is also required for the import of glutathione-S-conjugates into the yeast vacuole.[113] Notably, Ycf1 expression is indeed regulated by stress response factors such as Yap1[114,115], since *yap1* mutants exhibit cadmium hypersensitivity.[116]

Interestingly, in the fission yeast *S. pombe*, the 830-residue Hmt1 half-size ABC transporter of the vacuolar membrane can mediate both heavy metal resistance and peptide translocation.[117,118] Hmt1 is most closely related to P-glycoproteins from *Leishmania* implicated in heavy metal resistance of the parasites.[119] Recently, Hmt1 has been demonstrated to be an ATP-dependent transporter of phytochelatins, which are metal-binding, peptide-Cd^{2+} complexes.[118] Thus, like Ycf1, Hmt1 mediates heavy metal tolerance through sequestration of metal ions into the vacuole. Taken together, these data demonstrate that yeast ABC transporters can indeed mediate transport of peptides across the vacuolar membrane.[113,118] However, it is not known whether Ycf1 or other yeast ABC transporters play a role in Ape1 or Ams1 translocation. The future analysis of mutants blocking the cytoplasm to vacuole transport of aminopeptidase I[110] may help to elucidate the molecular identity of this translocation machinery and verify a possible role of ABC transporters in the biogenesis of the vacuole.

B. MITOCHONDRIA AND ENDOPLASMIC RETICULUM

Most proteins that are imported from the cytoplasm into the mitochondria contain a N-terminal signal sequence. The corresponding protein import machinery has been well studied[20,22,120] and will therefore not be discussed here. Protein import of most mitochondrial polypeptides is thought to occur at contact sites between inner and outer membrane (IM and OM, respectively) requiring ATP-hydrolysis in the mitochondrial matrix.[20,22,121] ABC transporters have also been speculated to play a role in the biogenesis and assembly of mitochondria and chloroplasts.[7,122] Indeed, chloroplasts of *Marchantia polymorpha* contain a ABC protein, MbpX, but its function there is unknown.[123]

A search for novel yeast ABC genes required for mitochondrial biogenesis and eventually for protein import into mitochondria was recently initiated. PCR-cloning allowed for the isolation of at least ten different gene fragments encoding various ABC proteins.[124] While some sequences were corresponding to novel ABC sequences,[124] others turned out to be identical to already cloned ABC genes, including Adp1[125], Ssh1/Mdl2[86,126] and Mdl1.[126] One newly identified ABC protein gene, *atm1* (for ABC transporter of mitochondria), was indeed found to encode a mitochondrial protein.[124] Deletion of the *atm1* gene results in a slow-growth phenotype on rich medium and cessation of growth on minimal media. Moreover, a complete absence of mitochondrial cytochromes is observed, although a defective heme import could be ruled out as a possible cause.[124] Atm1 is a 694-residue TMS_6-ABC half-size ABC transporter. Based on cell fractionation and immunofluorescence experiments, Atm1 localizes to the inner mitochondrial membrane, with the ABC domain facing the mitochondrial matrix.[124] The inner membrane localization of Atm1 could be consistent with a function for Atm1 in mitochondrial protein import, since one can speculate that Atm1 could provide structural and/or energy-providing component of the import site. The architecture of such a pore is likely to form a hydrophilic, proteinacous channel across both mitochondrial membranes to allow for protein import to occur. Notably, puromycin, a known substrate for mammalian P-glycoproteins,[13] is a potent inhibitor of protein import, as it inhibits ATP-consumption in the matrix.[127] Similar to the mitochondrial protein import, it was proposed that the bacterial HlyB/HlyD transporter for hemolysin forms a protein pore stretching across both bacterial membranes at the membrane junctions,[76] although in bacterial hemolysin export the direction of protein transport is the opposite of mitochondrial import. At present it is not known whether Δatm1 mitochondria have a defective protein import machinery or if the slow-growth phenotype of Δatm1 cells is the result of another yet unidentified transport function of Atm1.

An intriguing alternative possibility would be that Atm1 is responsible for export of substrates from the mitochondrial matrix rather than import.[124] From the observation that mammalian P-glycoproteins can act as phospholipid-specific flippases[13] and steroid transporters,[98] one could perhaps speculate about a possible function for Atm1 in intramitochondrial lipid transport or lipid homeostasis. Certain membrane phospholipids such as phosphatidylserine and cardiolipin are either exclusively decarboxylated or localized in the mitochondria, respectively.[128,129] Thus, Atm1 could play a role in membrane lipid homeostasis at contact sites of the mitochondrial membranes.[130] For instance, cardiolipin depletion of the IM must be prevented, to ensure proper mitochondrial function. This process is particularly important at the contact sites where lipid translocation and/or intra-mitochondrial membrane lipid exchange is suspected to occur.[130] Further, decarboxylation of phosphatidylserine to phosphatidyethanolamine occurs exclusively in the IM. In turn, phosphatidylethanolamine acts as a precursor for phosphatidylcholine biosynthesis in the ER membrane and must therefore be exported from the inner mitochondrial membrane,[128] presumably via the membrane junctions of OM and IM.[130] While a function of Atm1 in these lipid translocation processes has not been established, it should be easy to test this idea in the near future.

In mammalian cells, antigen presentation of viral peptide antigens requires the action of a heterodimeric Tap1/Tap2 ABC transporter of the ER membrane (see also chapter 4). Recently, we have co-expressed human *tap1* and *tap2* in yeast, leading to a fully functional, heterodimeric Tap1/Tap2 peptide transporter localized to the endoplasmic reticulum (S. Urlinger et al, submitted for publication). Both Tap1 and Tap2 seem to carry sufficient ER targeting information and require no additional factors for proper function as a peptide transporter in the yeast ER as determined by in vitro peptide transport assays. Most interestingly, peptide transport studies in vitro using purified yeast microsomal vesicles with or without the human Tap1/Tap2 complex, revealed an additional novel peptide-specific and ATP-dependent transport activity in the ER. This novel peptide transport system in the ER is distinct from the classical ER protein translocation machinery[131-133] and independent of the Tap1/Tap2 complex (S. Urlinger et al, submitted for publication). In addition, the novel ER peptide transporter seems to mediate peptide uptake into the ER lumen rather than efflux. Thus, it is likely to be different from a previously characterized ER peptide transport activity, which was shown to mediate export of glycosylated tripeptides from the ER lumen into the cytoplasm.[134] While the existence of a yeast counterpart of the mammalian Tap1/Tap2 ABC transporter seems imaginable, the mechanism and function of such a peptide transport system of the yeast ER are currently unknown and its dissection will await further studies.

C. PEROXISOME

Peroxisome biogenesis involves the direct protein import into the organelle across the single lipid bilayer.[23-25] The capacity of peroxisomes to translocate protein precursors with targeting signals across the single membrane is well established.[23-25] Known peroxisomal targeting signals include PTS1, the carboxy-terminal consensus tripeptide SKL, and PTS2, the N-terminal consensus sequence RLX$_5$HL (where X is any amino acid). The PTS1 motif works in various organisms including yeast,[135] mammals, plants, and insects.[23] Interestingly, there is evidence that stably folded proteins[26] and even oligomeric enzymes[27] containing the PTS1 sequence are translocated into peroxisomes, suggesting that protein unfolding is not a pre-requisite for peroxisomal protein import. The PTS2 motif is sufficient to direct proteins, which are otherwise localized in the cytoplasm, to the peroxisomal matrix.[136] Mutations in the PTS2 consensus sequence of peroxisomal proteins can block their import into the organelle.[137]

In fungi, several peroxisomal assembly deficient (*pas*) mutants have been identified.[138] For instance, the *pas10* gene product was found to recognize proteins with the PTS1 targeting signal.[139] The functional Pas10 analog in the methylotrophic yeast *Pichia pastoris*, Pas8, is also a PTS1-specific receptor.[140] The actual import receptor for the PTS2 targeting signal was recently identified in the *pas7* gene product.[141] Other components of the peroxisomal import system and translocation mechanisms for proteins from the cytoplasm into the peroxisomal matrix have not been identified. Hence, the actual mechanism by which peroxisomal protein import is accomplished remains an enigma.

In humans, the lack of functional peroxisomes or a defective peroxisome import machinery has lethal consequences. For instance, mutations in the half-size ABC transporters Pmp70[142] and ALDp lead to Zellweger syndrome and adrenoleukodystrophy.[142,143] These fatal genetic diseases are associated with apparent defects in peroxisome function and/or organelle assembly, and with the demyelination of the nervous system that is paralleled by the accumulation of very-long-chain fatty acids in peroxisomes.[143-146] The physiological functions of Pmp70 and ALDp are still unknown, but it has been suggested that these ABC proteins may operate as protein or lipid transporters in the peroxisomal membrane.[142] For instance, the removal of Pmp70 from the cytosolic face of the peroxisomal membrane by limited proteolysis in vitro leads to a severely impaired import of acyl-CoA oxidase.[142] Therefore, the authors suggested a speculative role for Pmp70, direct or indirect, in peroxisomal protein import.

In the yeast *Saccharomyces cerevisiae*, ABC transporters are also involved and required for peroxisome biogenesis. Two yeast ABC proteins genes, *ssh1* and *ssh2* (for *s*terile *s*ix *h*omologues), both of which are closely related to mammalian P-glycoproteins, were originally identified by PCR-cloning.[86] After cloning the full-length gene, Ssh2 was renamed Pal1 (for *p*eroxisomal *A*BC transporter-*l*ike protein).[147] *ssh1* is identical to *mdl2* (for *Mdr*-*l*ike), which was independently isolated using a similar approach.[126] In addition, the same group has cloned another ABC transporter, Mdl1, that is closely related but different to both Ssh1/Mdl2 and Ssh2.[126]

The half-size ABC protein Pal1 is the closest homolog of the human peroxisomal ABC transporters Pmp70[142,143,145] and ALDp,[146] implicated in the peroxisomal disorders such as Zellweger syndrome and X-linked adrenoleukodystrophy, respectively. Interestingly enough, yeast *Δpal1* cells, despite being viable under normal growth conditions, are unable to grow on oleate as the sole carbon source.[147] Growth of yeast cells on oleate requires functional peroxisomes, and the Pal1 transcript is only detectable in oleate-grown cells, suggesting Pal1 to be a constituent of the yeast peroxisomal membrane. *pal1* is identical to *pxa1* which was independently cloned by another group.[148] Pal1/Pxa1 and another closely related gene, YKL741, which was identified during the systematic yeast genome sequencing project,[149] form heterodimers to comprise a functional ABC transporter in the peroxisomal membrane.[147] Deletion of YKL741, which was renamed Pxa2,[150] resulted in a growth phenotype identical to that of Pal1/Pxa1.[148] Thus, as proposed for mammalian Pmp70 and/or ALPp, yeast Ssh2/Pal1/Pxa1 and Pxa2 could play a role in peroxisome biogenesis and/or organelle proliferation.

Recently, missense mutations were introduced into several regions of *pxa1* that are conserved among ABC transporters such as Pmp70, ALDp, Tap1 and several P-glycoproteins.[150] Two motifs, designated loop1 and EAA-like, which in bacterial ABC transporters were suggested to predict substrate specificity, were shown to be important for Pxa1 function. Interestingly, within the EAA-like motif, there is a glutamic acid residue (E291 in ALDp) conserved among all peroxisomal transporters. Mutations of E291 in the human ALPp transporter were found in several unrelated patients suffering from adrenoleukodystrophy. In future work, it will be important to identify the natural substrates of the human peroxisomal ABC transporters and to determine if mammalian Pmp70 and ALDp, when functionally expressed in yeast, are able to restore growth of *Δpal1* cells on oleate as the sole carbon source. The first direct demonstration that yeast Pal1 is a transporter for lipid substrates comes from a very recent report in which the authors show (and unfortunately introduce yet another new and confusing nomenclature for Pal1 and Pxa2, namely Pat2 and Pat1) that a heterodimeric Pat1-Pat2 ABC transporter is required for the import of long-chain fatty acids into peroxisomes of *Saccharomyces cerevisiae.*[151]

IV. CONCLUSION AND PERSPECTIVES

The lower eukaryote yeast has become an evolutionary storehouse for numerous ABC proteins, several of which appear to have structural and/or functional homologs in mammalian cells. However, with few exceptions, for the majority of the sequenced yeast ABC transporters, in vivo substrates and functions and their subcellular localization are presently unknown (see Table 2.1).

While overexpression of several yeast ABC transporters is linked to multidrug resistance phenomena,[16,17,59,152,153] others including Ste6, Hmt1, and Ycf1 clearly exhibit peptide transport activity.[10,11,112,113,117,118] The most prominent among them is Ste6,[9] whose in vivo function is to mediate the nonclassical export of the a-factor mating pheromone. In addition, certain yeast multidrug resistance transporters such as Pdr5 and Snq2, while otherwise involved in pleiotropic drug resistance (PDR) development in yeast[59,154] and part of the yeast PDR network,[155] were shown to mediate transport of steroid substrates[156] and even peptides.[18]

It seems clear that peptide, protein and lipid translocation must occur during biogenesis of all cellular organelles. However, the components or the transport machineries as well as in vivo substrates are still unknown. There are several distinct yeast ABC protein genes (Table 2.1), which are good candidates for cellular functions such as peptide transport across membranes, vesicle-independent secretion and organelle biogenesis.[59] The future biochemical characterization of these yeast ABC transporters, especially the effect of deletion mutants of ABC transporter genes on organelle biogenesis, and the reconstitution of their transport activities in vitro, will certainly help to shed light on their speculative role in organelle biogenesis.

The accessibility of yeast genetics makes this lower eukaryote a valuable model system for the molecular analysis of both endogenous and heterologous ABC proteins. In bacteria, the purification of individual ABC proteins and their functional reconstitution into proteo-liposomes has substantially contributed to our understanding of how these bacterial ABC transport systems work in vivo. The yeast Ste6 A-factor transporter represents an good candidate protein for similar studies because its physiological substrate is known. Likewise, other yeast ABC transporters, which appear to be potential structural and functional homologs of mammalian ABC transporters implicated in genetic disease, are also excellent candidates for in vitro studies to elucidate their possible in vivo substrates. Similar studies can be carried out on mammalian ABC proteins, since they can be functionally expressed in yeast.[87,89,90,93,157] Furthermore, in cases where a functional complementation between yeast and animal ABC transporters has been demonstrated, the construction of chimeric ABC transporters can be utilized for a molecular-genetic structure-function analysis to pinpoint domains of functional importance in ABC proteins from different organisms.[158]

Moreover, the discovery of yeast ABC protein gene homologs of mammalian genes implicated in the pathology of human disease has further strengthened the importance of yeast as a model system for molecular-genetic structure-function studies on eukaryotic ABC transporters. Yeast genetics enables the isolation and characterization of transport mutants of any ABC transporter, since expression of mutant ABC proteins in yeast can be monitored by phenotypic selection and/or other functional assays. Future work will have to focus on extensive structural analysis, including crystallization attempts of individual ABC transporters to allow for uncovering the molecular mechanism of ABC protein function.

ACKNOWLEDGMENTS

We wish to thank colleagues and collaborators, in particular P. Bissinger, P. Chambon, A. Delahodde, A. Goffeau, C. Jacq, S. Kane, R. Kölling, N. Kralli, Y. Lemoine, R. Losson, W. Nichols, P. Piper, J. Rine, D. Sanglard, J. Subik, J. Thorner and D. Wolf for their generosity in sharing research materials, yeast strains and reagents. Thanks also for many fruitful discussions and privileged communications of unpublished data.

Current research work in my laboratory is supported by grants from the Austrian Science Foundation (P-MOB-10123 and FWF-SFB-604), the Austrian National Bank (OENB project 5638), by funds from the "Herzfelder'schen Familienstiftung" and in part by NIH grant #RO1-CA64645-01A1. R. E. is a recipient of a postdoctoral fellowship from the Deutsche Forschungsgemeinschaft.

REFERENCES

1. Walter P, Johnson AE. Signal sequence recognition and protein targeting to the endoplasmic reticulum membrane. Ann Rev Cell Biol 1994; 10: 87-119.

2. von Heijne G. Signal sequences: The limits of variation. J Mol Biol 1985; 184: 99-105.
3. Rothman JE, Wieland FT. Protein sorting by transport vesicles. Science 1996; 272: 227-234.
4. Schekman R, Orci L. Coat proteins and vesicle budding. Science 1996; 271: 1526-1533.
5. Raymond CK, Roberts CJ, Moore KE et al. Biogenesis of the vacuole in *Saccharomyces cerevisiae*. Int Rev Cytol 1992; 139: 59-120.
6. Stack JH, Emr SD. Genetic and biochemical studies of protein sorting to the yeast vacuole. Curr Opin Cell Biol 1993; 5: 641-646.
7. Kuchler K, Thorner J. Secretion of peptides and proteins lacking hydrophobic signal sequences: The role of ATP-driven membrane translocators. Endocrine Rev 1992; 13: 499-514.
8. Kuchler K. Unusual routes of protein secretion: the easy way out. Trends Cell Biol 1993; 3: 421-426.
9. Kuchler K, Swartzman ER, Thorner J. A novel mechanism for transmembrane translocation of peptides: the *Saccharomyces cerevisiae* Ste6 transporter and export of the mating pheromone a-factor. Curr Top Membranes 1994; 41: 19-41.
10. McGrath JP, Varshavsky A. The yeast *STE6* gene encodes a homologue of the mammalian multidrug resistance P-glycoprotein. Nature 1989; 340: 400-404.
11. Kuchler K, Sterne RS, Thorner J. *Saccharomyces cerevisae STE6* gene product: a novel pathway for protein export in eukaryotic cells. EMBO J 1989; 8: 3973-3984.
12. Higgins CF. ABC-transporters: from microorganisms to man. Ann Rev Cell Biol 1992; 8: 67-113.
13. Kane SE. Multidrug resistance of cancer cells. In: Testa B, Meyer, UA, eds. *Advances in Drug Research*. San Diego: Academic Press Ltd., 1996; 28: 181-252.
14. Gottesman MM, Hrycyna CA, Schoenlein PV et al. Genetic analysis of the multidrug transporter. Ann Rev Genet 1995; 29: 607-649.
15. Higgins CF. The ABC of channel regulation. Cell 1995; 82: 693-696.
16. Bissinger P, Kuchler K. Molecular cloning and expression of the *Saccharomyces cerevisiae STS1* gene product: a yeast ABC-transporter conferring mycotoxin resistance. J Biol Chem 1994; 269: 4180-4186.
17. Balzi E, Wang M, Leterme S et al. *PDR5*, a novel yeast multidrug resistance conferring transporter controlled by the transcription regulator *PDR1*. J Biol Chem 1994; 269: 2206-2214.
18. Kolaczkowski M, van der Rest M, Cybularz-Kolaczkowska A et al. Anticancer drugs, ionophoric peptides, and steroids as substrates of the yeast multidrug transporter Pdr5p. J Biol Chem 1996; 271: 31543-31548.
19. Hill A, Ploegh H. Getting inside out: The transporter associated with antigen processing (TAP) and the presentation of viral antigen. Proc Natl Acad Sci USA 1995; 92: 341-343.

20. Schatz G, Dobberstein B. Common principles of protein translocation across membranes. Science 1996; 271: 1519-1526.
21. Schatz G. The protein import system of mitochondria. J Biol Chem 1996; 271: 31763-31766.
22. Stuart RA, Neupert W. Topogenesis of inner membrane proteins of mitochondria. Trends Biochem Sci 1996; 21: 261-267.
23. Subramani S. Protein import into peroxisomes and biogenesis of the organelle. Ann Rev Cell Biol 1993; 9: 445-478.
24. Kunau WH, Hartig A. Peroxisome biogenesis in *Saccharomyces cerevisiae*. Antonie Van Leeuwenhoek 1992; 62: 63-78.
25. Aitchison JD, Nuttley WM, Szilard AM et al. Peroxisome biogenesis in yeast. Mol Microbiol 1992; 6: 3455-3460.
26. Walton PA, Hill PE, Subramani S. Import of stably folded proteins into peroxisomes. Mol Biol Cell 1995; 6: 675-683.
27. Mc New JA, Goodman JM. An oligomeric protein is imported into peroxisomes in vivo. J Cell Biol 1994; 127: 1245-1257.
28. Yoshihisa T, Anraku Y. A novel pathway of import of α-mannosidase, a marker enzyme of vacuolar membrane, in *Saccharomyces cerevisiae*. J Biol Chem 1990; 265: 22418-22425.
29. Klionsky DJ, Cueva R, Yaver DS. Aminopeptidase I of *Saccharomyces cerevisiae* is localized to the vacuole independent of the secretory pathway. J Cell Biol 1992; 119: 287-299.
30. Segui-Real B, Martinez M, Sandoval IV. Yeast aminopeptidase I is post-translationally sorted from the cytosol to the vacuole by a mechanism mediated by its bipartite N-terminal extension. EMBO J 1995; 14: 5476-5484.
31. Oda MN, Scott SV, Hefner-Gravink A et al. Identification of a cytoplasm to vacuole targeting determinant in aminopeptidase I. J Cell Biol 1996; 132: 999-1010.
32. Schultz J, Ferguson B, Sprague Jr GF. Signal transduction and growth control in yeast. Curr Opin Gen Dev 1995; 5: 31-37.
33. Sprague Jr GF, Thorner J. Pheromone response and signal transduction during the mating process of *Saccharomyces cerevisiae*. In: Broach JR, Pringle JR, Jones EW, eds. The Molecular Biology of the Yeast *Saccharomyces cerevisiae*, Second Edition. Cold Spring Harbor: Cold Spring Harbor Laboratory Press, 1992: 657-744
34. Brake AJ, Brenner C, Najarian R et al. Structure of genes encoding precursors of the yeast peptide mating pheromone a-factor. In: Gething M-J, ed. Protein Transport and Secretion. New York: Cold Spring Harbor Laboratory Press, 1985: 103-108.
35. Adames N, Blundell K, Ashby MN et al. Role of yeast insulin-degrading enzyme homologs in pro-pheromone processing and bud site selection. Science 1995; 270: 464-467.
36. Ashby MN, King DS, Rine J. Endoproteolytic processing of a farnesylated peptide in vitro. Proc Natl Acad Sci USA 1992; 89: 4613-4617.

37. Hrycyna CA, Clarke S. Maturation of isoprenylated proteins in *Saccharomyces cerevisiae*. Multiple activities catalyze the cleavage of the three carboxy-terminal amino acids from farnesylated substrates in vitro. J Biol Chem 1992; 267: 10457-10464.

38. Anderegg RJ, Betz R, Carr SA et al. Structure of the *Saccharomyces cerevisiae* mating hormone a-factor: identification of S-farnesyl cysteine as a structural component. J Biol Chem 1988; 263: 18236-18240.

39. Schafer W, Rine J. Protein prenylation: genes, enzymes, targets, and functions. Ann Rev Genet 1992; 26: 209-237.

40. Ashby MN, Rine J. Ras and a-factor converting enzyme. Methods Enzymol 1995; 250: 235-251.

41. Fujita A, Oka C, Arikawa Y et al. A yeast gene necessary for bud-site selection encodes a protein similar to insulin-degrading enzymes. Nature 1994; 372: 567-570.

42. He B, Chen P, Chen SY et al. RAM2, an essential gene of yeast, and *RAM1* encode the two polypeptide components of the farnesyltransferase that prenylates a-factor and Ras proteins. Proc Natl Acad Sci USA 1991; 88: 11373-11377.

43. Goodman LE, Judd SR, Farnsworth CC et al. Mutants of *Saccharomyces cerevisiae* defective in the farnesylation of Ras proteins. Proc Natl Acad Sci USA 1990; 87: 9665-9669.

44. Sterne-Marr RE, Blair LC, Thorner J. *Saccharomyces cerevisiae STE14* gene is required for COOH-terminal methylation of a-factor mating pheromone. J Biol Chem 1990; 265: 20057-20060.

45. Sapperstein S, Berkower C, Michaelis S. Nucleotide sequence of the yeast *STE14* gene, which encodes farnesyl cysteine carboxyl methyltransferase, and demonstration of its essential role in a-factor export. Mol Cell Biol 1994; 14: 1438-1449.

46. Hrycyna CA, Sapperstein SK, Clarke S et al. The *Saccharomyces cerevisiae STE14* gene encodes a methyltransferase that mediates C-terminal methylation of a-factor and RAS proteins. EMBO J 1991; 10: 1699-1709.

47. Ashby MN, Errada PR, Boyartchuk VL et al. Isolation and DNA sequence of the *STE14* gene encoding farnesyl cysteine carboxyl methyltransferase. Yeast 1993; 9: 907-913.

48. Gaynor EC, te Heesen S, Graham TR et al. Signal-mediated retrieval of a membrane protein from the Golgi to the ER in yeast. J Cell Biol 1994; 127: 653-665.

49. Letourneur F, Gaynor EC, Hennecke S et al. Coatomer is essential for retrieval of dilysine-tagged proteins to the endoplasmic reticulum. Cell 1994; 79: 1199-1207.

50. Townsley FM, Pelham HR. The KKXX signal mediates retrieval of membrane proteins from the Golgi to the ER in yeast. Eur J Cell Biol 1994; 64: 211-216.

51. Sterne RE. A novel pathway for peptide hormone biogenesis: processing and secretion of the mating pheromone a-factor by *Saccharomyces cerevisiae*. Ph.D. Thesis. University of California, Berkeley, 1989.

52. Kuchler K, Dohlman H, Thorner J. The a-factor transporter (*STE6* gene product) and cell polarity in *Saccharomyces cerevisiae*. J Cell Biol 1993; 120: 1203-1215.
53. Kölling R, Hollenberg CP. The ABC-transporter Ste6 accumulates in the plasma membrane in a ubiquitinated form in endocytosis mutants. EMBO J 1994; 13: 3261-3271.
54. Berkower C, Loayza D, Michaelis S. Metabolic instability and constitutive endocytosis of *STE6*, the a-factor transporter of *Saccharomyces cerevisiae*. Mol Biol Cell 1994; 3: 633-654.
55. Marcus S, Caldwell GA, Xue CB et al. Total in vitro maturation of the *Saccharomyces cerevisiae* a-factor lipopeptide mating pheromone. Biochem Biophys Res Commun 1990; 172: 1310-1316.
56. Berkower C, Michaelis S. Mutational analysis of the yeast a-factor transporter *STE6*, a member of the ATP binding cassette (ABC) protein superfamily. EMBO J 1991; 10: 3777-3785.
57. Browne BL, McClendon V, Bedwell DM. Mutations within the first LSGGQ motif of Ste6p cause defects in a-factor transport and mating in *Saccharomyces cerevisiae*. J Bacteriol 1996; 178: 1712-1719.
58. Epand RF, Xue CB, Wang SH et al. Role of prenylation in the interaction of the a-factor mating pheromone with phospholipid bilayers. Biochemistry 1993; 32: 8368-8373.
59. Egner R, Mahé Y, Pandjaitan R et al. ATP binding cassette transporters in yeast: from mating to multidrug resistance. In: Rothman S, ed. Membrane Protein Transport. Greenwich: JAI Press Inc., 1995; 2: 57-96.
60. Caldwell GA, Wang SH, Naider F et al. Consequences of altered isoprenylation targets on a-factor export and bioactivity. Proc Natl Acad Sci USA 1994; 91: 1275-1279.
61. Borst P, Schinkel AH, Smit JJ et al. Classical and novel forms of multidrug resistance and the physiological functions of P-glycoproteins in mammals. Pharmacol Ther 1993; 60: 289-299.
62. Hancock JF, Cadwallader K, Paterson H et al. A CAAX or a CAAL motif and a second signal are sufficient for plasma membrane targeting of ras proteins. EMBO J 1991; 10: 4033-4039.
63. Thissen JA, Casey PJ. Microsomal membranes contain a high affinity binding site for prenylated peptides. J Biol Chem 1993; 268: 13780-13783.
64. Michaelis S. STE6, the yeast a-factor transporter. Semin Cell Biol 1993; 4: 17-27.
65. Dinarello CA. The biology of interleukin-1. Chem Immunol 1992; 51: 1-32.
66. Dinarello CA. Role of interleukin-1 in infectious diseases. Immunol Rev 1992; 127: 119-146.
67. Rubartelli A, Cozzolino F, Talio M et al. A novel secretory pathway for Interleukin-1β, a protein lacking a signal sequence. EMBO J 1990; 9: 1503-1510.
68. Baldari C, Murray JA, Ghiara P et al. A novel leader peptide which al-

lows efficient secretion of a fragment of human interleukin-1β in *Saccharomyces cerevisiae.* EMBO J 1987; 6: 229-234.

69. Michaelis S, Herskowitz I. The a-factor pheromone of *Saccharomyces cerevisiae* is essential for mating. Mol Cell Biol 1988; 8: 1309-1318.

70. Hancock JF, Cadwallader K, Marshall CJ. Methylation and proteolysis are essential for efficient membrane binding of prenylated p21K-ras(B). EMBO J 1991; 10: 641-646.

71. Marcus S, Caldwell GA, Miller D et al. Significance of C-terminal cysteine modifications to the biological activity of the *Saccharomyces cerevisiae* a-factor mating pheromone. Mol Cell Biol 1991; 11: 3603-3612.

72. Caldwell GA, Wang SH, Xue CB et al. Molecular determinants of bioactivity of the *Saccharomyces cerevisiae* lipopeptide mating pheromone. J Biol Chem 1994; 269: 19817-19825.

73. Xue CB, Caldwell GA, Becker JM et al. Total synthesis of the lipopeptide a-mating factor of *Saccharomyces cerevisiae.* Biochem Biophys Res Commun 1989; 162: 253-257.

74. Elia L, Marsh L. Role of the ABC transporter Ste6 in cell fusion during yeast conjugation. J Cell Biol 1996; 135: 741-751.

75. Geller D, Taglicht D, Rotem E et al. Comperative topology studies in *Saccharomyces cerevisiae* and in *Escherichia coli.* J Biol Chem 1996; 271: 13746-13753.

76. Koronakis V, Hughes C. Bacterial signal peptide-independent protein export: HlyB-directed secretion of haemolysin. Semin Cell Biol 1993; 4: 7-15.

77. Meyer TH, vanEndert PM, Uebel S et al. Functional expression and purification of the ABC transporter complex associated with antigen processing (TAP) in insect cells. FEBS Lett 1994; 351: 443-447.

78. Madden K, Costigan C, Snyder M. Cell polarity and morphogenesis in *Saccharomyces cerevisiae.* Trend Cell Biol 1992; 2: 22-29.

79. Jackson CL, Hartwell LH. Courtship in *Saccharomyces cerevisiae*: both cell types choose mating partners by responding to the strongest pheromone signal. Cell 1990; 63: 1039-1051.

80. Davis NG, Horecka JL, Sprague Jr GF. *cis*-and *trans*-acting functions required for endocytosis of the yeast pheromone Receptors. J Cell Biol 1993; 122: 53-65.

81. Rohrer J, Benedetti H, Zanolari B et al. Identification of a novel sequence mediating regulated endocytosis of the G protein-coupled alpha-pheromone receptor in yeast. Mol Biol Cell 1993; 4: 511-521.

82. Hicke L, Riezman H. Ubiquitination of a yeast plasma membrane receptor signals its ligand-stimulated endocytosis. 1996; 84: 277-287.

83. Roth AF, Davis NG. Ubiquitination of the yeast a-factor receptor. J Cell Biol 1996; 134: 661-674.

84. Raths S, Rohrer J, Crausaz F et al. *end3* and *end4*: Two mutants defective in receptor-mediated and fluid-phase endocytosis in *Saccharomyces cerevisiae.* J Cell Biol 1993; 120: 55-65.

85. Jackson CL, Konopka JB, Hartwell LH. *S. cerevisiae* α pheromone receptors activate a novel signal transduction pathway for mating partner discrimination. Cell 1991; 67: 389-402.

86. Kuchler K, Göransson M, Visnawathan M et al. Dedicated transporters for peptide export and intercompartemental traffic in yeast. CSH Symp Quant Biol 1992; 57: 579-592.

87. Raymond M, Gros P, Whiteway M et al. Functional complementation of yeast *ste6* by a mammalian multidrug resistance *mdr* gene. Science 1992; 256: 232-234.

88. Cole SP, Bhardwaj G, Gerlach JH et al. Overexpression of a transporter gene in a multidrug-resistant human lung cancer cell line. Science 1992; 258: 1650-1654.

89. Volkman SK, Cowman AF, Wirth DF. Functional complementation of the *ste6* gene of *Saccharomyces cerevisiae* with the *pfmdr1* gene of *Plasmodium falciparum*. Proc Natl Acad Sci USA 1995; 92: 8921-8925.

90. Ruetz S, Brault M, Kast C et al. Functional expression of the multidrug resistance-associated protein in the yeast *Saccharomyces cerevisiae*. J Biol Chem 1996; 271: 4154-4160.

91. Toppmeyer DL, Slapak CA, Croop J et al. Role of P-glycoprotein in dolastatin 10 resistance. Biochem Pharmacol 1994; 48: 609-612.

92. Sharma RC, Inoue S, Roitelman J et al. Peptide transport by the multidrug resistance pump. J Biol Chem 1992; 267: 5731-5734.

93. Kuchler K, Thorner J. Functional expression of human *mdr1* cDNA in *Saccharomyces cerevisiae*. Proc Natl Acad Sci USA 1992; 89: 2302-2306.

94. Sharom FJ, DiDiodato G, Yu X et al. Interaction of the P-glycoprotein multidrug transporter with peptides and ionophores. J Biol Chem 1995; 270: 10334-10341.

95. van Helvoort A, Smith AJ, Sprong H et al. *MDR1* P-glycoprotein is a lipid translocase of broad specificity, while *MDR3* P-glycoprotein specifically translocates phosphatidylcholine. Cell 1996; 87: 507-517.

96. Smit JJM, Schinkel AH, Oude Elferink RPJ et al. Homozygous disruption of the murine *mdr2* P-glycoprotein gene leads to a complete absence of phospholipid from bile and to liver disease. Cell 1993; 75: 451-462.

97. Ruetz S, Gros P. Phosphatidylcholine translocase: a physiological role for the *mdr2* gene. Cell 1994; 77: 1071-81.

98. Ueda K, Okamura N, Hirai M et al. Human P-glycoprotein transports cortisol, aldosterone, and dexamethasone, but not progesterone. J Biol Chem 1992; 267: 24248-24252.

99. Muesch A, Hartmann E, Rohde K et al. A novel pathway for secretory proteins? Trends Biochem Sci 1990; 15: 86-88.

100. Rubartelli A, Bajetto A, Allavena G et al. Secretion of thioredoxin by normal and neoplastic cells through a leaderless secretory pathway. J Biol Chem 1992; 267: 24161-24164.

101. D'Amore PA. Modes of FGF release *in vivo* and *in vitro*. Cancer Metastasis Rev 1990; 9: 227-238.

102. Cooper DNW, Barondes SH. Evidence for export of a muscle lectin from cytosol to extracellular matrix and for a novel secretory mechanism. J Cell Biol 1990; 110: 1681-1691.

103. Cleves AE, Cooper DNW, Barondes SH et al. A new pathway for protein export in *Saccharomyces cerevisiae*. J Cell Biol 1996; 133: 1017-1026.

104. Hochstrasser M. Protein degradation or regulation: Ub the judge. Cell 1996; 84: 813-815.

105. Riezman H. Yeast endocytosis. Trends in Cell Biol 1993; 3: 273-277.

106. Takeshige K, Baba M, Tsuboi S et al. Autophagy in yeast demonstrated with proteinase-deficient mutants and conditions for its induction. J Cell Biol 1992; 119: 301-311.

107. Thumm M, Egner R, Koch B et al. Isolation of autophagocytosis mutants of *Saccharomyces cerevisiae*. FEBS Lett 1994; 349: 275-280.

108. Chang YH, Smith JA. Molecular cloning and sequencing of genomic DNA encoding aminopeptidase I from *Saccharomyces cerevisiae*. J Biol Chem 1989; 264: 6979-6983.

109. Scott SV, Klionsky DJ. *In vitro* reconstitution of cytoplasm to protein targeting in yeast. J Cell Biol 1995; 131: 1727-1735.

110. Harding TM, Morano KA, Scott SV et al. Isolation and characterisation of yeast mutants in the cytoplasm to vacuole protein targeting pathway. J Cell Biol 1995; 131: 591-602.

111. Harding TM, Hefner-Gravink A, Thumm M et al. Genetic and phenotypic overlap between autophagy and the cytoplasm to vacuole protein targeting pathway. J Biol Chem 1996; 271: 17621-17624.

112. Szczypka MS, Wemmie JA, Moye-Rowley WS et al. A yeast metal resistance protein similar to human cystic fibrosis transmembrane conductance regulator (CFTR) and multidrug resistance-associated protein. J Biol Chem 1994; 269: 22853-22857.

113. Li ZS, Szczypka M, Lu YP et al. The yeast cadmium factor protein (*YCF1*) is a vacuolar glutathione S-conjugate pump. J Biol Chem 1996; 271: 6509-6517.

114. Stephen DWS, Rivers SL, Jamieson DJ. The role of the *YAP1* and *YAP2* genes in the regulation of the adaptive oxidative stress responses of *Saccharomyces cerevisiae*. Mol Microbiol 1995; 16: 415-423.

115. Gounalaki N, Thireos G. Yap1p, a yeast transcriptional activator that mediates multidrug resistance, regulates the metabolic stress response. EMBO J 1994; 13: 4036-4041.

116. Wemmie JA, Wu AL, Harshman KD et al. Transcriptional activation mediated by the yeast AP-1 protein is required for normal cadmium tolerance. J Biol Chem 1994; 269: 14690-14697.

117. Ortiz DF, Kreppel L, Speiser DM et al. Heavy metal tolerance in the fission yeast requires an ATP-binding cassette-type vacuolar membrane transporter. EMBO J 1992; 11: 3491-3499.

118. Ortiz DF, Ruscitti T, McCue KF et al. Transport of metal-binding peptides by *HMT1*, a fission yeast ABC-type vacuolar membrane protein. J Biol Chem 1995; 270: 4721-4728.

119. Callahan HL, Beverley SM. Heavy metal resistance: a new role for P-glycoproteins in *Leishmania.* J Biol Chem 1991; 266: 18427-18430.

120. Glick B, Schatz G. Import of proteins into mitochondria. Ann Rev Genet 1991; 25: 21-44.

121. Pfanner N, Rassow J, van der Klei IJ et al. A dynamic model of the mitochondrial protein import machinery. Cell 1992; 68: 999-1002.

122. Kuchler K, Thorner J. Membrane translocation of proteins without hydrophobic signal sequences. Curr Opin Cell Biol 1990; 2: 617-624.

123. Umesono K, Inokuchi H, Shiki Y et al. Structure and organization of *Marchantia polymorpha* chloroplast genome. II. Gene organization of the large single copy region from *rps'12* to *atpB.* J Mol Biol 1988; 203: 299-331.

124. Leighton J, Schatz G. An ABC Transporter of the inner mitochondrial membrane is required for normal growth of yeast. EMBO J 1995; 14: 188-195.

125. Purnelle B, Skala J, Goffeau A. The product of the *YCR105* gene located on the chromosome III from *Saccharomyces cerevisiae* presents homologies to ATP-dependent permeases. Yeast 1991; 7: 867-872.

126. Dean M, Allikmets R, Gerrard B et al. Mapping and sequencing of two yeast genes belonging to the ATP-binding cassette superfamily. Yeast 1994; 10: 377-383.

127. Price J, Verner K. Puromycin inhibits protein import into mitochondria by interfering with an intramitochondrial ATP-dependent reaction. Biochim Biophys Acta 1993; 1150: 89-97.

128. Achleitner G, Zweytick D, Trotter PJ et al. Synthesis and intracellular transport of aminoglycerophospholipids in permeabilized cells of the yeast *Saccharomyces cerevisiae.* J Biol Chem 1995; 270: 29836-29842.

129. Kent C. Eukaryotic phospholipid biosynthesis. Ann Rev Biochem 1995; 64: 315-343.

130. Simbeni R, Pon L, Zinser E et al. Mitochondrial membrane contact sites of yeast. Characterization of lipid components and possible involvement in intramitochondrial translocation of phospholipids. J Biol Chem 1991; 266: 10047-10049.

131. Sanders SL, Schekman R. Polypeptide translocation across the endoplasmic reticulum membrane. J Biol Chem 1992; 267: 13791-13794.

132. Sanders SL, Whitfield KM, Vogel JP et al. Sec61p and BiP directly facilitate polypeptide translocation into the ER. Cell 1992; 69: 353-365.

133. Brodsky JL, Goeckeler J, Schekman R. BiP and Sec63p are required for both co- and post-translational protein translocation into the yeast endoplasmic reticulum. Proc Natl Acad Sci USA 1995; 92: 9643-9646.

134. Römisch K, Schekman R. Distinct processes mediate glycoprotein and glycopeptide export from the endoplasmic reticulum in *Saccharomyces cerevisiae.* Proc Natl Acad Sci USA 1992; 89: 7227-7231.

135. Kragler F, Langeder A, Raupachova J et al. Two independent peroxisomal targeting signals in catalase A of *Saccharomyces cerevisiae.* J Cell Biol 1993; 120: 665-673.

136. Erdmann R. The peroxisomal targeting signal of 3-oxoacyl-CoA thiolase from *Saccharomyces cerevisiae*. Yeast 1994; 10: 935-944.

137. Glover JR, Andrews DW, Subramani S et al. Mutagenesis of the amino targeting signal of *Saccharomyces cerevisiae* 3-ketoacyl-CoA thiolase reveals conserved amino acids required for import into peroxisomes *in vivo*. J Biol Chem 1994; 269: 7558-7563.

138. Kunau WH, Beyer A, Franken T et al. Two complementary approaches to study peroxisome biogenesis in *Saccharomyces cerevisiae*: forward and reversed genetics. Biochimie 1993; 75: 209-224.

139. Van der Leij I, Franse MM, Elgersma Y et al. *PAS10* is a tetratricopeptide-repeat protein that is essential for the import of most matrix proteins into peroxisomes of *Saccharomyces cerevisiae*. Proc Natl Acad Sci USA 1993; 90: 11782-11786.

140. Terlecky SR, Nuttley WM, McCollum D et al. The *Pichia pastoris* peroxisomal protein Pas8p is the receptor for the C-terminal tripeptide peroxisomal targeting signal. EMBO J 1995; 14: 3627-3634.

141. Rehling P, Marzioch M, Niesen F et al. The import receptor for the peroxisomal targeting signal 2 (PTS2) in *Saccharomyces cerevisiae* is encoded by the *PAS7* gene. EMBO J 1996; 15: 2901-2913.

142. Kamijo K, Taketani S, Yokota S et al. The 70-kDa peroxisomal membrane protein is a member of the Mdr (P-glycoprotein)-related ATP-binding protein superfamily. J Biol Chem 1990; 265: 4534-4540.

143. Gartner J, Moser H, Vallee D. Mutations in the 70 kd peroxisomal membrane protein gene in Zellweger syndrome. Nature Gen 1992; 1: 16-20.

144. Aubourg P, Mosser J, Douar AM et al. Adrenoleukodystrophy gene: unexpected homology to a protein involved in peroxisome biogenesis. Biochimie 1993; 75: 293-302.

145. Gartner J, Valle D. The 70 kDa peroxisomal membrane protein: an ATP-binding cassette transporter protein involved in peroxisome biogenesis. Semin Cell Biol 1993; 4: 45-52.

146. Mosser J, Douar A, Sarde C et al. Putative X-linked adrenoleukodystrophy gene shares unexpected homology with ABC-transporters. Nature 1993; 361: 726-730.

147. Swartzman EE, Viswanathan MN, Thorner J. The *PAL1* gene product is a peroxisomal ATP-binding cassette transporter in the yeast *Saccharomyces cerevisiae*. J Cell Biol 1996; 132: 549-563.

148. Shani N, Watkins PA, Valle DL. *PXA1*, a possible *Saccharomyces cerevisiae* ortholog of the human adrenolykodystrophy gene. Proc Natl Acad Sci USA 1995; 92: 6012-6016.

149. Bossier P, Fernandes L, Vilela C et al. The yeast *YKL741* gene situated on the left arm of chromosome XI codes for a homologue of the human ALD protein. Yeast 1994; 10: 681-686.

150. Shani N, Sapag A, Valle D. Characterization and analysis of conserved motifs in a peroxisomal ATP-binding cassette transporter. J Biol Chem 1996; 271: 8725-8730.

151. Hettema EH, van Roermund CWT, Distel B et al. The ABC transporter proteins Pat1 and Pat2 are required for import of long-chain fatty acids into peroxisomes of *Saccharomyces cerevisiae*. EMBO J 1996; 15: 3813-3822.

152. Servos J, Haase E, Brendel M. Gene *SNQ2* of *Saccharomyces cerevisiae*, which confers resistance to 4-nitroquinoline-N-oxide and other chemicals, encodes a 169 kDa protein homologous to ATP-dependent permeases. Mol Gen Genet 1993; 236: 214-218.

153. Katzmann DJ, Hallstrom TC, Voet M et al. Expression of an ATP-binding cassette transporter-encoding gene (*YOR1*) is required for oligomycin resistance in *Saccharomyces cerevisiae*. Mol Cell Biol 1995; 15: 6875-6883.

154. Balzi E, Goffeau A. Genetics and biochemistry of yeast multidrug resistance. Biochim Biophys Acta 1994; 1187: 152-162.

155. Mahé Y, Parle-McDermott, Nourani A et al. The ATP-binding cassette multidrug transporter Snq2 of *Saccharomyces cerevisiae*: a novel target for the transcription factors Pdr1 and Pdr3. Mol Microbiol 1996; 20: 109-117.

156. Mahé Y, Lemoine Y, Kuchler K. The ATP binding cassette transporters Pdr5 and Snq2 of *Saccharomyces cerevisiae* can mediate transport of steroids *in vivo*. J Biol Chem 1996; 271: 25167-25175.

157. Ueda K, Shimabuku AM, Konishi H et al. Functional expression of human P-glycoprotein in *Schizosaccharomyces pombe*. FEBS Lett 1993; 330: 279-282.

158. Teem JL, Berger HA, Ostedgaard LS et al. Identification of revertants for the cystic fibrosis delta F508 mutation using *STE6-CFTR* chimeras in yeast. Cell 1993; 73: 335-346.

159. Hirata D, Yano K, Miyahara K et al. *Saccharomyces cerevisiae YDR1*, which encodes a member of the ATP-binding cassette (ABC) superfamily, is required for multidrug resistance. Curr Genet 1994; 26: 285-294.

160. Kralli A, Bohen SP, Yamamoto KR. *LEM1*, an ATP-binding cassette transporter, selectively modulates the biological potency of steroid hormones. Proc Natl Acad Sci USA 1995; 92: 4701-4705.

161. Parle-McDermott AG, Hand NJ, Goulding SE et al. Sequence of 29 kb around the *PDR10* locus on the right arm of *Saccharomyces cerevisiae* chromosome XV: similarity to part of chromosome I. Yeast 1996; 12: 999-1004.

162. Decottignies A, Lambert L, Catty P et al. Identification and characterization of Snq2, a new multidrug ATP binding cassette transporter of the yeast plasma membrane. J Biol Chem 1995; 270: 18150-18157.

163. Sandbaken M, Lupisella JA, Di-Domenico B et al. Isolation and characterization of the structural gene encoding elongation factor 3. Biochim Biophys Acta 1990; 1050: 230-234.

164. Vazquez de Aldana CR, Marton MJ, Hinnebusch AG. *GCN20*, a novel ATP binding cassette protein, and *GCN1* reside in a complex that mediates activation of the eLF-2 alpha kinase *GCN2* in amino acid-starved cells. EMBO J 1995; 14: 3184-3199.

165. Nishi K, Yoshida M, Nishimura M et al. A leptomycin B resistance gene of *Schizosaccharomyces pombe* encodes a protein similar to the mammalian P-glycoproteins. Mol Microbiol 1992; 6: 761-769.
166. Turi TG, Rose JK. Characterization of a novel *Schizosaccharomyces pombe* multidrug resistance transporter conferring brefeldin A resistance. Biochem Biophys Res Commun 1995; 213: 410-418.
167. Prasad R, de Wergifosse P, Goffeau A et al. Molecular cloning and characterization of a novel multidrug resistance gene of *Candida albicans*, *CDR1*. Curr Gen 1995; 27: 320-329.
168. Sanglard D, Kuchler K, Ischer F et al. Mechanisms of resistance to azole antifungal agents in *Candida albicans* isolates from AIDS patients involve specific multidrug transporters. Antimicrob Agents Chemother 1995; 39: 2378-2386.
169. Sanglard D, Ischer F, Monod M et al. Cloning of *Candida albicans* genes conferring resistance to azole antifungal agents: characterization of *CDR2*, a new multidrug ABC transporter gene. Microbiology 1996; in press.

CHAPTER 3

SECRETION OF MAMMALIAN PROTEINS THAT LACK A SIGNAL SEQUENCE

Anna Rubartelli and Roberto Sitia

I. INTRODUCTION

Cells are surrounded by an impermeable membrane whose integrity is essential to maintain the differences (in ionic concentration, redox potential etc.) between the cytoplasm and the extracellular space. Yet, cells avoid isolation and communicate by sending each other macromolecular messages in the form of secretory proteins. These molecules, interacting with specific receptors that span the cell membrane, activate tightly regulated systems which transduce the signal to the nucleus and elicit the appropriate responses.

Secretory proteins begin their synthesis within the cytosol: thus, in order to get out of the cell they need to translocate through a membrane. The problem of how some proteins can be secreted overcoming the topological barrier of the cell membranes was solved in the early 1970s by the discovery of the signal sequence (or leader peptide) and the subsequent demonstration that translocation of secretory proteins occurs cotranslationally at the ER membrane.[1] The secretory signal sequence which directs the rest of the protein to and across the ER membrane is a hydrophobic 15-30 residues region, typically situated at the N-terminus of the nascent polypeptide that is removed from the growing polypeptide chain as soon as it translocates into the ER lumen.

Membrane translocation as an essential step to get a proper location is not limited to secretory proteins:[2] the synthesis of almost all proteins occurs in the cytosol, and hence also proteins destined to mitochondria, chloroplasts and peroxisomes have to overcome a membrane to reach their final destination (Fig. 3.1). The mechanisms underlying the movement from the cytosol into the different organelles have some common themes. Distinct classes of topogenic sequences within newly synthesized proteins are thought to interact with receptor proteins

Unusual Secretory Pathways: From Bacteria to Man, edited by Karl Kuchler, Anna Rubartelli and Barry Holland. © 1997 R.G. Landes Company.

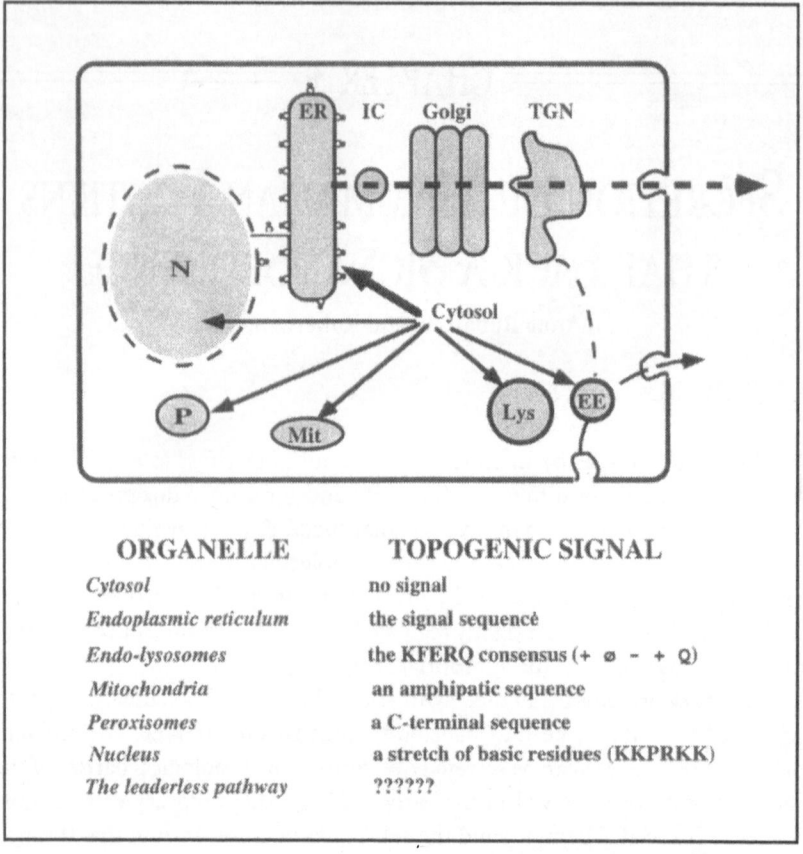

ORGANELLE	TOPOGENIC SIGNAL
Cytosol	no signal
Endoplasmic reticulum	the signal sequencé
Endo-lysosomes	the KFERQ consensus (+ ø - + Q)
Mitochondria	an amphipatic sequence
Peroxisomes	a C-terminal sequence
Nucleus	a stretch of basic residues (KKPRKK)
The leaderless pathway	??????

Fig. 3.1. Many proteins must cross a membrane to reach their final destination. The problem of crossing a membrane is not limited to secretory proteins. The membranes of peroxisomes, endolysosomal compartment and mitochondria are equipped with systems that selectively deliver proteins to their lumen. In the case of mitochondria, the problem is particularly complex owing to the existence of an inner membrane in these organelles.[3] The identification of the sequences that mediate the specific targeting to individual membranes (topogenic sequences) has allowed the construction of chimeric proteins with novel, predetermined localization.[104]

on the surface of the target organelles. The actual translocation of the target membrane is the less defined step: the main problems that remain to be solved are how the membrane remains impermeable while being crossed by molecules of different sizes and how vectoriality is determined (Fig. 3.2). Generally, translocation appears to be an energy-dependent process (Fig. 3.3). Proteins must be unfolded to translocate, and cellular chaperone molecules have been implicated (for a comprehensive review, see Schatz and Dobberstein).[3]

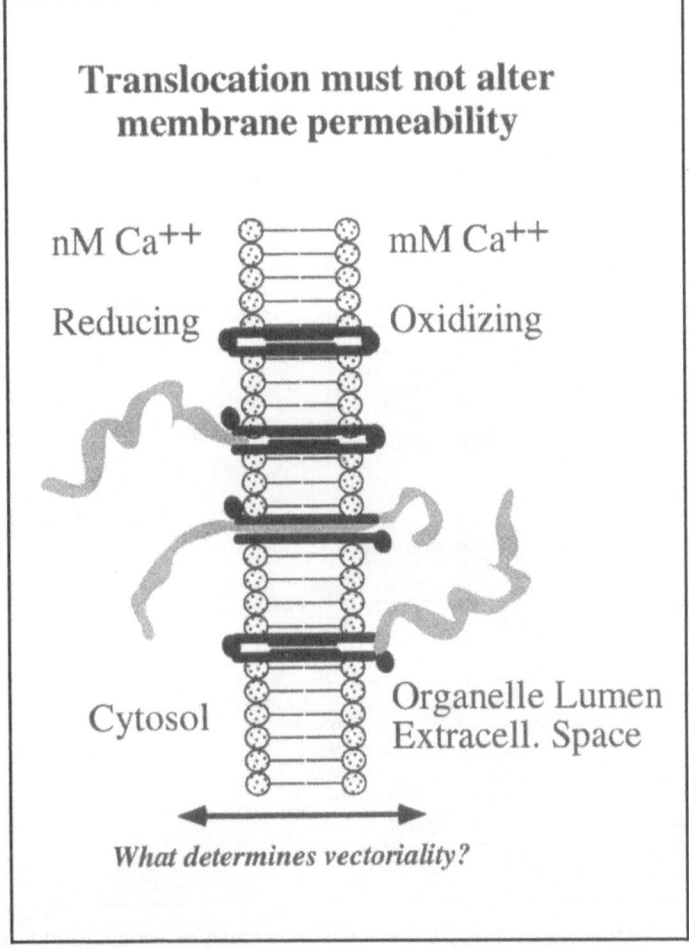

Translocation must not alter membrane permeability

nM Ca^{++} mM Ca^{++}

Reducing   Oxidizing

Cytosol   Organelle Lumen
Extracell. Space

What determines vectoriality?

Fig. 3.2. Translocation must not alter membrane permeability. One of the crucial problems that cells face in delivering proteins across a membrane is that of maintaining the ionic, pH and redox gradients between the cytosol and the lumena of the different organelles or the extracellular space. It follows that the opening of the 'translocons' (the molecular complexes responsible for translocation) must be tightly regulated, possibly by the binding of the cognate topogenic sequences.[105] Soluble proteins cross the membrane vectorially. The hydrophobic 'stop transfer' sequences that allow the insertion of integral membrane proteins are thought to be extruded laterally from the translocon.[3]

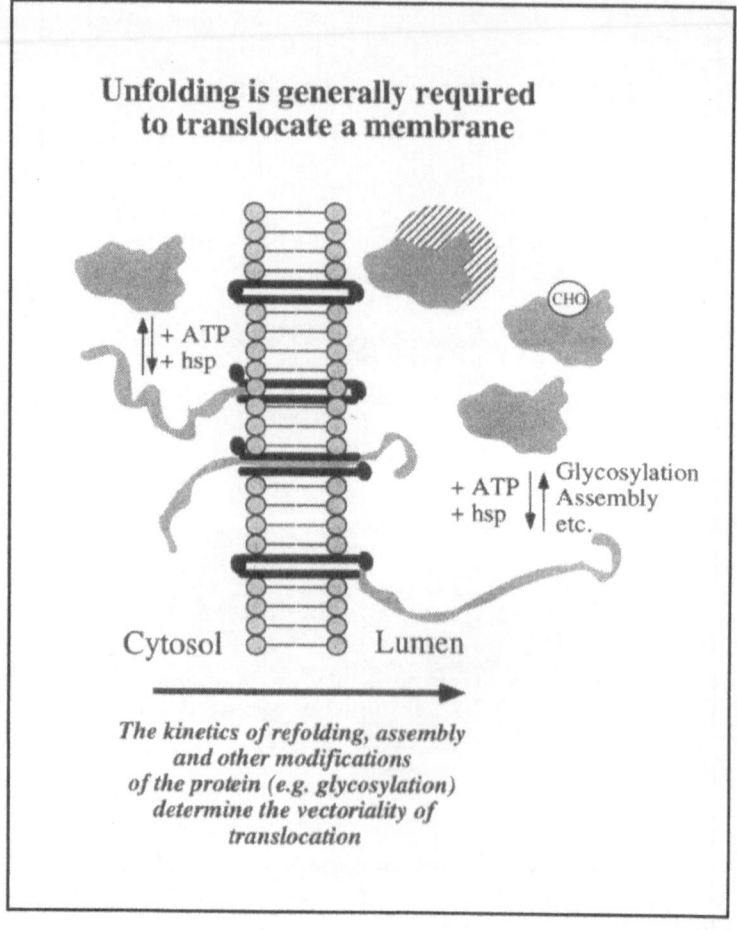

Fig. 3.3. How is vectoriality determined? Many lines of evidence indicate that proteins must be unfolded in order to translocate a membrane.[3] Any modification that reduces the diffusibility of the polypeptide across the membrane is likely to favor the accumulation on one side of the membrane. Thus, glycosylation, (re)folding or assembly with other macromolecules may determine the vectoriality of translocation.[106]

Recent studies indicate that the membranes delimiting the various compartments (more than a dozen in eukaryotic cells!) display striking structural differences: peripheral membranes (the prototype being the plasmamembrane, PM) are rich in cholesterol, ordered and thicker, while the central membranes (prototype: endoplasmic reticulum (ER) membrane)[4] are poor in cholesterol, disordered, thinner and possibly more permeable. The differences in composition may account for the different propensity of the various membranes to mediate translocation of macromolecules. Indeed, translocation across the PM has not been formally demonstrated so far. Sequencing of a few soluble media-tors, such as interleukin-1 (IL-1)-α and β, and fibroblast growth factors (FGFs), revealed the absence of a leader peptide,[5,6] raising the question of how these proteins can reach the extracellular space.[7] At first it was thought that cell death was responsible for the release of these 'leaderless' proteins; but a simple "kami-kaze" model is not consistent with the observations that leaderless polypeptide secretion is selective[8] and can result in autocrine stimulation.[9] Genetic evidence from studies in *E.coli* and yeast confirmed that some polypeptides leave the cells through alternative pathways that involve ATP-binding transporters.[10] Whether similar transporters are responsible for leaderless secretion in higher eukaryotes remains to be investigated, but it seems clear that an alternative secretory pathway exists.

Since the pionieristic observations on IL-1 and FGF, several secretory pro-teins which are devoid of a typical hydrophobic signal peptide have been described (Table 3.1).

Leaderless secretory proteins in mammals display a relatively low molecu-lar mass, which ranges from 12 to 45 kDa for most of them, with few excep-tions, such as proteins of the transglutaminases family, that have molecular weights of 70-80 kDa. Common features of these proteins are the absence of N-linked oligosaccharides in spite of many potential N-glycosylation acceptor sites and the presence of free cysteines not involved in disulfide bridges. Fur-thermore, some of them are N-acylated. These properties suggest that these proteins do not enter the ER lumen.

How do these leaderless proteins reach the extracellular space? Do they uti-lize a single pathway, or does more than one way exist to leave cells without going through the exocytic pathway? What are the advantages given by the al-ternative pathway(s)? So far there are not definitive answers to these crucial questions.

Before discussing the possible mechanisms that underlie the secretion of leaderless proteins we shall summarize some of the best known examples.

II. LEADERLESS PROTEINS SECRETED BY MAMMALIAN CELLS

INTERLEUKIN-1

Two forms of IL-1 exist, IL-1α and β, that are encoded by two different genes but share the same cell surface receptors.[11] IL-1 is a mediator of inflammation and plays a central role in the regulation of immune responses. This cytokine is produced primarily in response to infection, endotoxin and inflammatory agents, but it is also involved in regulating cell growth and/or differentiation under normal[11] and pathological[9] conditions. Both IL-1 α and β are synthesized

Table 3.1. Mammalian leaderless secretory proteins

Protein	Reference
Interleukin-1	8
Interleukin-1 Receptor Antagonist	23,24
Basic Fibroblast Growth Factor, bFGF (FGF-2)	30,31
Acidic Fibroblast Growth Factor, aFGF (FGF-1)	32,33
Glia Activating Factor (FGF-9)	29
Ciliary NeuroTrophic Factor	91
Mammary-Derived Growth Inhibitor	87
Platelet-Derived Endothelial Cell Growth Factor	92,93
Thioredoxin/ADF	46,50
Endothelial-Monocyte Activating Polypeptide II	94
Annexin 1/Lipocortin 1	95, 96
Prothymosine	97
Parathymosine	98
Coagulation Factor XIIIa	99
Transglutaminase (tissue)	100,101
Transglutaminase (prostate)	84
Lectin L14/galectin-I	60
Mac-2/L29/CBP30	55,56,63,64
β-Galactoside Binding Protein	58
HMG1	102
HIV1-tat	67
Rhodanese	103
Interleukin-1-Converting Enzyme?	17,18,21

This is not meant as a comprehensive list of leaderless secretory proteins. For instance, not all cases are listed where proteins have been characterized in more than one species. Moreover, as more proteins belonging to this class are being characterized, the list will be probably incomplete at the time of publication. We apologize to those scientists whose work has not been duly acknowledged and cited.

as 33 kDa intracellular precursors that are subsequently cleaved into mature proteins of 17 kDa. While pro-IL-1α is biologically active, the precursor of IL-1β is not. Most studies on IL-1 secretion have focused on IL-1β, as this cytokine is the major extracellular form.

The problem of how IL-1, lacking a signal peptide, can be secreted by the producing cells was at first set aside, with the assumption that this protein is released by cell lysis. Hence, it was thought that monocytes would produce abundant IL-1 and release it by some sort of suicide (apoptosis was not yet in fashion in those days). The hypothesis of cell death as a mechanism of secretion was subsequently ruled out by several observations:

(1) the 33 kDa pro-IL-1β is not generally released in the extracellular space, where only the 17 kDa form is found.

(2) IL-1β secretion is selective, as its extracellular appearance does not correlate with the presence of cytoplasmic markers such as, for instance, lactate dehydrogenase.[8]

(3) IL-1β and α are secreted with different kinetics[12,8] and there are drugs that block IL-1β secretion, but not the release of classical secretory proteins.[8]

That IL-1β indeed follows an unconventional secretory route was established by culturing activated monocytes in the presence of drugs (such as brefeldin A and monensin) which inhibit transport within the classical secretory pathway.[8] Under these conditions, secretion of IL-1β is not blocked, but actually stimulated.

Besides drugs inhibiting the classical secretion, other apparently unrelated drugs increase IL-1β secretion to different extents.[8] These include cycloheximide, an inhibitor of protein synthesis, DNP and CCCP, two uncouplers of oxidative phosphorylation, and the calcium ionophore A23187.

One of the strongest inducers of IL-1β secretion is ATP. At first, IL-1β secretion observed following treatment with extracellular ATP was proposed to be due to apoptosis.[13] However, while activation of IL-1β secretion possibly occurs during the first phases of apoptosis, apoptosis per se does not seem necessary for the efficient release of bioactive IL-1β.[14] A more recent report suggests that ATP actually stimulates IL-1β secretion as a consequence of increasing the activity of the IL-1β converting enzyme (ICE) by depletion of intracellular K^+.[15]

Indeed, there is some controversy as to whether processing is required for IL-1β secretion or whether it is only cotemporal: while it is clear hat 17 kDa IL-1β is secreted to a greater extent than the precursor,[8,16] some reports have showed that a shift to secretion of the 33 kDa precursor occurs under certain culture conditions, such as a raise of extracellular pH[14] or treatment with ICE inhibitors.[17]

At any rate, the effect of extracellular ATP is not restricted to IL-1β maturation, as secretion of another leaderless protein, annexin 1, which does not undergo proteolytic maturation, is similarly increased by this drug (Fig. 3.4, panels A and B). As a control, exogenous ATP has no effect on the rate of secretion of a classical secretory protein, IL-6 (Fig. 3.4, panels C).

An unsolved problem in IL-1β maturation is where and how processing occurs. The cysteine protease (ICE) responsible for cleaving the 33 kDa to the mature form[17-18] is part of a novel family of proteins, some members of which are involved in regulation of apoptosis.[19] ICE is a heterodimeric protease composed of two subunits, p20 and p10. Mature ICE is derived from a 45 kDa proenzyme by proteolytic excision of an 11 kDa N-terminal precursor domain and of an internal sequence of 19 AA residues. Both the precursor and the active ICE are found in the cytosolic fraction,[20] while only the mature form has been

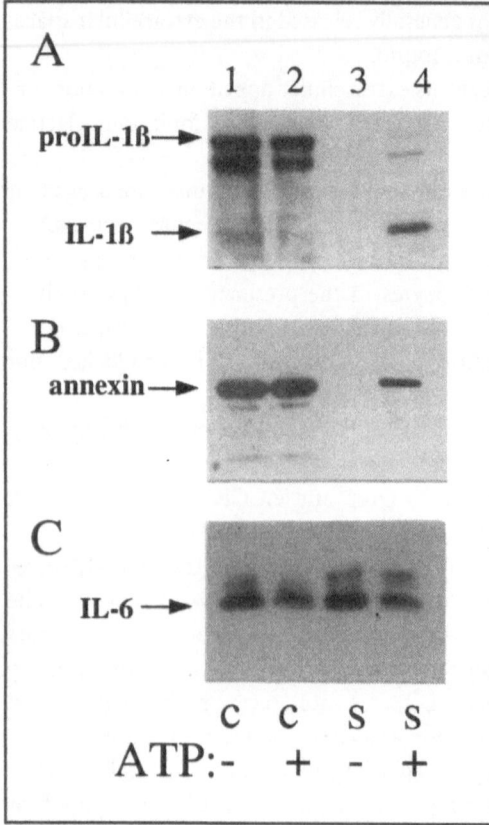

Fig. 3.4. ATP induces secretion of IL-1β and annexin 1. Human monocytes, activated with lipopolysaccharide for 1 h, were cultured 30 min in the presence or absence of 1 mM ATP. At the end of the incubation, supernatants were harvested, cells were lysed. Aliquots of supernatants and cell lysates were resolved by SDS-polyacrylamide gel electrophoresis and analyzed by western blotting with anti- IL-1β (panel A), anti-annexin 1 (panel B) or anti-IL-6 (panel C). While secretion of IL-6 is unaffected by ATP, both secretion of IL-1 and annexin 1 is induced by ATP. Anti-IL-1β and anti-IL-6 antibodies were kind gifts of Dr. Dinarello, anti-annexin 1[96] was a kind gift of Dr. Solito.

reported to localize on the external cell surface membrane.[21] As also ICE is devoid of a signal sequence, the problem of its externalization falls in the general issue of the secretion of leaderless proteins.

Activation of ICE precursor is observed after depletion of cellular K^+.[15,22] Several mechanisms may lower K^+ concentration in intact cells. A simple one may be the fusion of Na^+ rich endosomes with intracellular vesicles. Interestingly, some cellular IL-1β is present in trypsin resistant vesicles,[8] with some features of endosomes (AR, unpublished results). Cytosolic ICE, colocalized with pro-IL-1β within the cytosol, where [K^+] is high, is inactive and thus unable to process IL-1β: this would explain why intracellular 17 kDa IL-1β is never detected in the cytosol. A fraction of pro-IL-1β and ICE might then translocate in a subcompartment with low K^+, where activation of ICE and IL-1β conversion takes place just prior to secretion.

IL-1 function is controlled through a large number of regulatory circuits which involve, in addition to pre- and posttranslational regulatory events, the production of specific IL-1 inhibitors. The latter act as receptor antagonists (IL-1ra).[11] In monocytes IL-1ra is produced mostly in a form endowed with a signal peptide, and hence secreted through the classical pathway.[11] Interestingly, a structural variant of IL-1ra which lacks a leader sequence is abundantly produced by keratinocytes,[23] cells which express constitutively also a large amount of IL-1. This form accumulates mostly in the cytoplasm, albeit a small amount is released through a pathway insensible to brefeldin A.[24] The presence of an intracellular receptor antagonist suggest that also IL-1 may have some intracellular roles possibly involving an intracellular form of IL-1 receptor.

FIBROBLAST GROWTH FACTOR (FGF)

FGFs are members of a family of proteins involved in the regulation of cell proliferation, differentiation and function. They play important roles in normal development, tissue homeostasis, wound healing and repair (for reviews, see refs. 25-26). These proteins have also been shown to have roles in several pathological conditions, such as tumorigenesis and metastasis. They act on cells of various origins, through interaction with high and low affinity surface receptors. At least 5 high affinity receptors have been discovered; some of them bind and are activated by multiple FGFs, while others have a higher degree of specificity. FGF-receptor interaction is complicated by the role of low affinity receptors. Indeed, FGFs bind to polyanions. Binding to heparan sulphates of the cell surface or of the extracellular matrix seems necessary for the interaction with basic FGF to its high affinity membrane-bound receptor.[27]

Nine members of the FGF family have been identified, displaying 30-50% sequence homology. Of these, FGF-1 (or acidic FGF, AMW 17 kDa),[28] FGF-2 (or basic FGF, AMW 17 kDa)[6] and FGF-9 (or Glia Activating Factor, GAF, AMW 30 kDa)[29] are devoid of a signal peptide.

Experiments with single cells revealed that secretion of FGF-2 influences the migration of the very cell that secreted it, thus formally ruling out the possibility of a release due to cell death.[30] The same drugs that potentiate IL-1 secretion (such as exocytosis inhibitors, see above) also enhance secretion of FGF-2, indicating that FGF-2 is secreted through a pathway alternative to the classical one, which shares some features with the IL-1β secretory route. [31]

In the case of FGF-1, a different mechanism of secretion has been proposed.[32-33] FGF-1 (unlike FGF-2) is released by transfected cells in response to heat shock; also in this case, secretion is increased by treatment with brefeldin A. Interestingly, a cysteine residue is critical for the release of FGF-1 in response to heat shock. The wild type molecule, but not a mutant which lacks cys 30, is secreted as a latent homodimer with low affinity for heparin. Activation by reducing agents seems to be necessary to generate heparin binding and biological activity. Furthermore, the C-terminal region of FGF-1 seems able to bind to phosphatidylserine, suggesting that a step in FGF-1 export might involve FGF-1 targeting to phospholipids of the plasmamembrane or of other intracellular vesicles.

Both FGF-1 and FGF-2 are found in considerable extent in the nucleus of the cells that produce them.[34-35] Although a clear function for nuclear FGF has not been identified, a nuclear localization of endogenous FGF-1 and 2 is not surprising, as both of them are endowed with a putative nuclear localization sequence.

More intriguing is the reported ability of exogenous FGFs of reaching the nucleus of living cells from the extracellular space.[36-37] Translocation of exogenous FGF-1 into the nucleus, in addition to binding to high affinity surface receptor, seems necessary for achieving a full stimulation of target cells.[37] This ability is peculiar, as cells are surrounded by the plasma membrane, that acts as a hydrophobic barrier to most macromolecules and should not in principle allow the entry of exogenous proteins. However, it is not unique, as some bacterial and plant toxins and, more recently, other leaderless secretory proteins (such as IL-1 and HIV-Tat) turned out to be able to cross the cell membrane from the extracellular space and reach the cytoplasm or the nucleus of target cells.[38] The finding that other members of the family of leaderless secretory proteins reach the nucleus of target cells raises the possibility that these proteins utilize the same mechanism(s) to cross the membrane both in the outward and in the inward direction.

FGF-9 or glia activating factor is still more intriguing: in spite of lacking a classical secretory signal sequence, it displays N-linked glycosylation. This suggests that FGF-9 can go through the classical ER/Golgi secretion pathway. Indeed, secretion of FGF-9 seems to be much more efficient than secretion of other leaderless proteins.[29] Further studies are needed to understand whether or not this protein indeed translocates into the ER lumen, possibly via an internal signal sequence. In this connection, it is worth to recall that an internal cryptic uncleaved secretory signal sequence was originally identified in chicken ovalbumin; this protein translocates posttranslationally into the ER. Translocation is rather inefficient and a fraction of ovalbumin remains within the cytosol.[39] Furthermore, it has been recently demonstrated that under certain in vitro conditions, polypeptides that do not contain signal peptides may be mistranslocated into the ER.[40]

The other members of the FGF family are readily secreted proteins, that bear a conventional signal sequence. Interestingly, they have transforming potential, and some of them were initially identified as oncogenes. Clinically, FGF5 has been suggested to contribute to the possible autocrine growth of

AIDS-related Kaposi sarcomas.[41] The property of inducing transformation was observed for FGF-2 and FGF-1 only under rather contriving conditions. For instance, FGF-2 was reported to transform a fibroblast cell line only when transfected in an engineered form bearing a secretory leader sequence.[42] These observations indicate that transformation occurs more easily when the ligand and its receptor meet intracellularly, and suggest that the unconventional secretion of FGF-1 and FGF-2 might have evolved to avoid intracellular compartmentalization with their specific receptors. They also indicate that the exposure of FGF to the conditions of the ER lumen does not inhibit biological activity (see below).

THIOREDOXIN

Thioredoxin (TRX), a disulfide-reducing dithiol enzyme of 13 kDa, serves many intracellular biological functions as a hydrogen donor.[43,44] In accord with its main cytosolic localization, thioredoxin lacks any kind of known targeting or signal sequence.[45] In spite of this, however, thioredoxin has been isolated from the supernatants of a T cell leukemia as an autocrine growth factor, and has been termed adult T cell leukemia-derived factor (ADF).[46] Exogenous thioredoxin was found to exert cytokine activities, such as induction of cell proliferation in neoplastic T and B lymphocytes, possibly via upregulation of interleukin-2 receptor expression;[46-47] furthermore, it turned out to be identical to eosinophil cytotoxicity enhancing factor[48] and to a B cell hybridoma derived factor, able to induce proliferation and differentiation of B-chronic lymphocytic leukemia cells.[49] Thus, TRX belongs to the family of leaderless secretory proteins and, indeed, it is actively secreted through a leaderless secretory pathway that shares many features with that described for IL-1, such as the slow kinetics of secretion, the enhancing effect on secretion of several unrelated drugs and the sensitivity to methylamine.[50] Despite these similarities, while secretion of IL-1β is restricted to a few cell types, normal or neoplastic cells of many origins are able to secrete TRX. However, a cell specificity exists also for secretion of TRX, in that not all the cells that synthesize TRX also secrete it, even within the same lineage. For instance, secretion of thioredoxin is developmentally regulated in normal B and T lymphocytes, being more abundant in activated than in resting lymphocytes.[50,51] Furthermore, studies on normal and malignant hepatocytes revealed that TRX is synthetized by both normal liver cells and the hepatocarcinoma cell line HepG2; only the former, however, release abundant thioredoxin extracellularly.[52] When cultured in mild reducing conditions HepG2 cells, but not normal hepatocytes, increase the rate of thioredoxin secretion and undergo growth inhibition. Thus, secretion of thioredoxin by liver cells correlates with growth inhibition: normal, nonproliferating hepatocytes constitutively secrete thioredoxin, while the rapidly dividing HepG2 cells synthetize similar amounts of this enzyme but retain most of it intracellularly.[52]

Also exogenous recombinant thioredoxin inhibits proliferation of HepG2 cells, although to a lesser extent. In contrast, exogenous thiols and thioredoxin stimulate proliferation of a B cell lymphoma line, indicating that different cell types respond differently to variations in the extracellular redox potential.[52] Also in this experimental system, 2ME is more effective than thioredoxin. This may be due to the fact that, although reduced before addition, exogenous thioredoxin can rapidly be oxidized when diluted in the culture medium. How can secreted thioredoxin remain active in the oxidizing extracellular milieu? It is tempting to speculate that the concomitant secretion of sulphydryl compounds (such as glutathione or cysteine) might both maintain TRX reduced and stimulate its secretion. A similar mechanism might regulate the activity of FGF-1. Furthermore, in autocrine/paracrine systems, the "presentation" of the growth factor by the producing to the target cell may take place in a specialized microenvironment (reducing in the case of thioredoxin) which can favor the endogenous factor over that provided exogenously. In this connection, a membrane associated thioredoxin has been reported.[53]

EXTRACELLULAR LECTINS (L14 , L29/MAC-2, MGBP)

Lectins are carbohydrate-binding proteins that have two special properties: specificity for particular sugar residues and bivalency or polyvalency. In spite of being leaderless, some soluble lectins are present at extracellular sites in the developing or adult tissues that make them. Some of them are concentrated around cell clusters, as in the extracellular matrix, while others are at the interface between cells and the external milieu, such as in mucin. The extracellular function of these lectins remains a subject of speculation. It has been proposed that they are involved in differentiation (shaping the extracellular environment through interactions with glycoconjugates);[54] inflammation (Mac2[55,56] is expressed at high levels by inflammatory macrophages) or cancer (increasing the metastatic potential, possibly by mediating cellular recognition and adhesion in embolization).[57] Mouse β-galactoside binding proteins turned out to be an autocrine negative growth factor, with cytostatic properties. The growth inhibitory effect is not related to lectin properties but is consistent with mechanisms involving ligand-receptor interactions.[58]

Another peculiarity of extracellular lectins is that some of them are homodimers, which require reducing agents to maintain their carbohydrate-binding activity. As already discussed for thioredoxin, it is unclear how they can remain active in the oxidizing extracellular milieu.

The first suggestion for the unexpected secretion of these lectins came from immunohistochemical studies. In some cases, these studies pointed to a shift from an intracellular to an extracellular location with differentiation. For example, chicken-lactose-lectin II was found concentrated intracellularly in developing chicken muscle but extracellularly upon maturation.[59] The export mechanism of its mammals homolog, L-14, was then studied in more detail.[60] The mechanism of release proposed for L-14 appears to be different from that described for other leaderless proteins. In undifferentiated myoblasts, L-14 is diffusely distributed throughout the cytosol. As cells progressively differentiate, cytosolic L-14 concentrates in the cytosol next to the plasmamembrane and then accumulates in restricted regions beneath the membrane; these area containing concentrated L-14 molecules are then included in protrusions of the plasmamembrane which are evaginated to form extracellular vesicles highly enriched in the lectin. The latter is then released in the extracellular environment upon degeneration of the vesicle membrane. The problems remain as to what determines the selective concentration of L-14 in specific cytosolic regions under the plasmamembrane as well as to what then drives their evagination. It would be interesting to investigate the lipid composition of the released vesicles, which seem permeable very soon after their release.

This export process is reminiscent of the mechanism by which differentiating erythrocytes get rid of transferrin receptors: these are selectively shed in released vesicles, called exosomes.[61] However, unlike L-14 containing vesicles, exosomes originate from multivescicular bodies; that is, they originate from invagination (rather than evagination) of specific areas of the plasmamembrane that are internalized into multivesicular bodies and then released upon fusion of the latter with the plasmamembrane (see Raposo et al, this volume). Membrane blebbing, with formation of membrane-bound extracellular vesicles, specifically enriched in certain cytosolic proteins have also been found in calcifying tissues, and called "matrix vesicles."[62]

Studies on the export of other soluble lectins, L29[63] and CBP30[64] provided the first evidence of polarized secretion of leaderless proteins. Both human L29 and BHK CBP30 are secreted by a nonclassical pathway preferentially from the apical membrane of filter grown epithelial cells.[63,64] Differently from L14, however, there is no evidence for blebbing vesicles enriched in these lectins, suggesting a different mechanism of secretion. Indeed, similarly to IL-1β in human monocytes,[8] CBP30 in BHK cells is both diffused within the cytosol and aggregated in vesicular structures. As the pharmacology of CBP30 secretion[64] is identical to that of IL-1β[8] (secretion is potentiated by drugs blocking ER-Golgi transport, by cycloheximide and calcium ionophores while is inhibited by methylamine), it is tempting to speculate that apical secretion of this lectin requires membrane translocation rather than membrane blebbing.

HIV1-TAT

The human immunodeficiency virus (HIV-1) encodes a small (86 AA) transcription factor, tat, which, in conjunction with cellular polymerases and transcription factors, is responsible for activation of viral gene expression. Disruption of the *tat* gene prevents viral replication, indicating its essential role in the HIV-1 life cycle. Unique features in the DNA and RNA regulatory regions of the HIV-1 LTR make it a target for *the Tat* protein.[65] In addition to its action on viral LTR, tat is also able to drive the transcription of a number of cellular genes, such as some cytokine genes.[66]

According to its nuclear localization and function, Tat is endowed with a nuclear localization sequence (NLS). However, Tat molecules have also been found extracellularly,[67] suggesting that a fraction of newly made Tat, both in infected or transfected cells, can escape the nuclear targeting and be secreted (see below, Fig. 3.6). Indeed, in addition to its nuclear role of transcription factor, Tat has been reported to exert many extracellular activities interfering with growth regulation of infected or uninfected target cells.[67-69]

Extracellular Tat can be taken up by other infected or noninfected cells, and reach their nucleus where it drives transcription of endogenous or reporter genes.[70,71] This behavior is not unique, but, as anticipated above, it is shared with some bacterial and plant toxins, as well as two leaderless cytokines (FGF and IL-1), and two proteins involved in transcriptional control (a homeobox peptide and lactoferrin), that are all able to cross the cell membrane from the extracellular space and reach the cytoplasm or the nucleus of target cells.[38]

III. MOLECULAR MECHANISMS

RECOGNITION

In most of the above cases, cell death as a mechanism of release has been excluded. In general, while the protein of interest is found in the medium, other cytosolic proteins are not. Particularly relevant in this context are the experiments of Mignatti et al[30] who exploited an ingenious protocol to show bFGF secretion at the single cell level. Hence, if cells do not burst open, leaderless proteins must be somehow recognized among myriads of cytosolic macromolecules: they must be marked by a targeting signal which would warrant their selective release. Yet, it has not been possible to evince common motifs within the sequences of these proteins. This obviously does not exclude that such motifs exist, and the search for them is still, and more, active.

A pathway for degradation of cytosolic proteins, which involves posttranslational translocation into endosomes and lysosomes has been described:[72,73] under conditions of serum starvation up to 30% of cytosolic proteins are translocated and rapidly degraded. A loose consensus sequence (the prototype being

"KFERQ") appears responsible for the targeting of cytosolic proteins to endolysosomes. Despite the fact that these proteins are normally degraded, extracellular release of fragments and even of the whole translocated protein has been documented. A putative receptor molecule for KFERQ containing proteins has been recently identified on the lysosomal membrane.[74] The mechanism of translocation to endolysosomes requires the presence of hsc73, which binds substrate proteins at the KFERQ-like region and facilitates their import. The activity of hsc73 in KFERQ-dependent translocation is stimulated by reducing agents.[73]

hsc73 is a member of the heat shock proteins. These cytosolic factors, formerly described as proteins induced by stress conditions, are required in many processes of posttranslational translocation of proteins to intracellular organelles such as mithocondria. The role of heat shock proteins appears to be that of maintaining the protein in an unfolded state, compatible for translocation.[75] Although a KFERQ consensus sequence is absent from many leaderless secretory proteins, due to the high degeneracy of targeting sequences, the presence of a cryptic sequence mediating the same translocation cannot be excluded.

Actually, the KFERQ-dependent endolysosomes transport shares some features with the leaderless secretion pathway: for instance, secretion of FGF-1 is induced by heat shock;[32,33] secretion of IL-1, FGF-2 and thioredoxin is increased by treatment of drugs known to induce stress proteins;[8,14,31,50] in cells that are competent for IL-1β secretion, a fraction of pro-IL-1β is found within intracellular vesicles;[8] secretion of thioredoxin is induced by reducing agents.[52]

Exit

Once recognized, proteins to be secreted must be transported to the extracellular space. This can be accomplished by either of two mechanisms: translocation or vesiculation (Fig. 3.5).

Translocation

Most proteins destined for intracellular organelles or for export must cross at least one membrane en route to their destination. Fully folded proteins are not competent for export in prokaryotes,[76] nor for import into mitochondria.[77] In the case of import into the ER lumen, translocation usually occurs cotranslationally: in the few examples of posttranslational translocation reported in the literature, folded polypeptides were incompetent for translocation.[78] Cytoplasmic chaperones are thought to maintain proteins in an unfolded translocation-competent state until they encounter their target membrane(s). Even this dogma, however, has its exceptions. In Gram-negative bacteria, proteins of the Type II pathway may fold in the periplasm prior to secretion (see de Lima Pimenta et al, this volume). In mammals, a peroxisomal protein was shown to

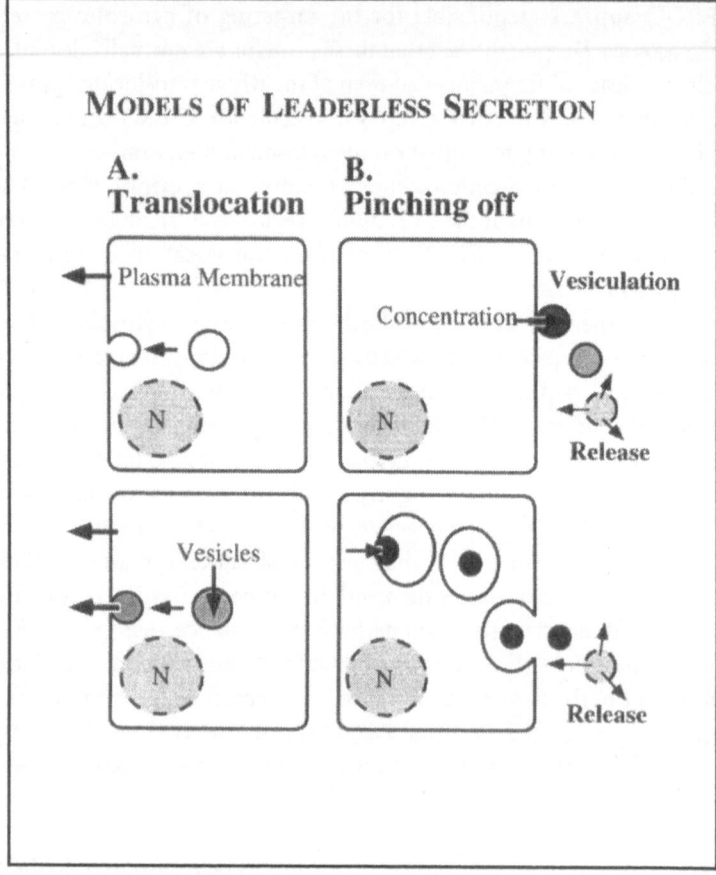

Fig. 3.5. Models of leaderless secretion. The selective release of certain cytosolic proteins can occur in at least two ways: translocation or concentration and vesiculation (pinching off). In both cases, mechanisms must exist by which the protein to be secreted is recognized. In the two left panels, the protein translocates directly at the plasmamembrane (upper panel) or within intracellular vesicles that then fuse with the plasmamembrane (lower panel).[8] Similarly, concentration in special regions of the cytosol beneath the membrane, evagination and release of the contents can occur at the plasmamembrane[60] or within multivesicular bodies (Raposo et al, this volume).

be imported into peroxisomes as a heterotrimer;[79] in plants, a fusion protein containing dihydrofolate reductase translocates within chloroplasts even when bound to methotrexate, a condition known to stabilize the folding of this enzyme.[80]

The requirement of unfolding for leaderless secretion has not been demonstrated so far. In some cases the intracellular leaderless protein is in an at least partially unfolded state. An example is the IL-1β precursor, which is known to be more protease sensitive than the mature form,[81] and is not biologically active.

Proteins that have a well defined intracellular function, however, such as thioredoxin,[43,44] annexin,[82] HIV-Tat,[65] are fully folded when in the cytosol. How then can they cross a membrane to reach the extracellular space? The presence of large pores able to accommodate folded proteins is unlikely. Not only have such structures not been observed in cellular membranes other than the nuclear envelope, which indeed allows import of large folded proteins,[83] but their presence could be incompatible with the cell viability, as the impermeability of the membrane would be jeopardized. Rather, one can speculate that these hypothetic pores have dynamic rather than static structures, their components being recruited and assembled after binding of the protein to the membrane to be translocated; after translocation, the pore would dissociate into subunits. The formation of large pores has been proposed also as a possible mechanism of import of a trimeric protein into peroxisomes.[79] Alternatively, unfolding by chaperones must be envisaged. If this is the case, how the molecules destined to secretion are selected among the other cytosolic molecules of a given leaderless secretory protein remains to be understood. According to these models, the slow kinetics of leaderless secretion could be due to the rate-limiting steps of pore formation or unfoldases activity.

Sequestration and vesiculation

It is to be stressed, however, that, at least in some cases, translocation of a membrane is dispensable in order to reach the extracellular space; fully folded molecules of a given protein may concentrate into patches beneath localized regions of the plasmamembrane which then evaginate and release extracellularly vesicles enriched in that protein, such as in the case of the lectin L14.[60] A similar mechanism is possibly involved in secretion of prostate transglutaminase:[84] immunogold electron microscopy of rodent prostate cells showed the presence of prostate transglutaminase in apocrine secretory vesicles that are pinched off from the apical plasmamembrane into the lumen.[85] While in this model the presence of unfoldases or large pores is dispensed, the problem remains as to how, as discussed above for lectins, the molecules that will be secreted are selected among the other molecules of a given leaderless protein and brought under the membrane, as well as to what induces the membrane to pinch off. Posttranslational modifications that increase the lipophylicity of proteins, such as acylation, myristoilation or farnesylation may favor their concentration in the proximity of the membrane to be crossed. Indeed, many leaderless secretory proteins undergo such modifications.[7]

VECTORIALITY AND POLARIZATION OF LEADERLESS SECRETION

The problem of how vectoriality is maintained in leaderless secretion is of interest, as many leaderless secretory proteins (FGF, HIV-tat, IL-1) seem to be also able to enter living cells and hence to translocate a membrane in both directions.[38] As depicted in Figure 3.3, folding and binding to compartmentalized molecules may be in competition with translocation, thereby favoring accumulation on one side of the membrane. This may be particularly relevant for proteins endowed with biological functions both inside and outside the cell, such as thioredoxin or Tat.

The secretion of a leaderless protein, the lectin L30, by epithelial cells is also polarized.[63,64] This implies that "translocons" responsible of secretion are themselves polarized. Another mechanism that would contribute to generating a local accumulation on one side of the secreting cell might be binding to polarized extracellular molecules.

IV. WHAT ARE THE FUNCTIONAL ROLES OF LEADERLESS SECRETION?

Leaderless secretory proteins have generally a large cytosolic pool, and their secretion is slow and inefficient compared to that of classical secretory proteins. Why then do cells utilize an alternative pathway of secretion? In some cases, leaderless secretion may actually reflect an intracellular function. For instance, thioredoxin is a well-known cytosolic oxide-reductase,[43,44] annexin 1 is involved in the control of exocytosis,[82] HIV-1 Tat is a transcription factor (Fig. 3.6).[65]

In autocrine systems, another advantage is the existence of different pathways of export for a ligand and its receptor which may allow efficient compartmentalization and/or independent posttranslational regulation[14] of the two molecules (Fig. 3.7). Indeed, transfection of a cell line bearing the FGF receptor with FGF-2 fused to a leader sequence resulted in cell transformation;[42] thus, intracellular localization of ligand and receptor might result in uncontrolled cell growth and neoplasia.

Interestingly, of the few soluble factors involved in growth inhibition so far identified, most are leaderless secretory proteins. They include mammary gland derived 13K polypeptide,[87] a soluble lectin of 15K,[58] IL-1 and thioredoxin which can both stimulate or inhibit proliferation depending on the cell type.[9,88,47,52]

Notably, exceptions are interferons and transforming growth factor-β, two growth inhibitory factors which are classical secretory proteins. However, interferon-γ mediates growth arrest of HeLa cells through a mechanism which implicates a role for thioredoxin, as shown by the finding that anti-sense thioredoxin makes Hela cells resistant to interferon-γ.[89] Thus, also in the case of leader sequence-bearing inhibitory growth factors, it is possible that a downstream component of the inhibitory autocrine loop is a leaderless secretory protein. However, the mechanisms underlying the inhibitory action of these negative growth factors are largely unknown.

Yet another good reason for following an alternative secretory pathway may be avoiding the oxidizing conditions or high calcium concentration of the ER

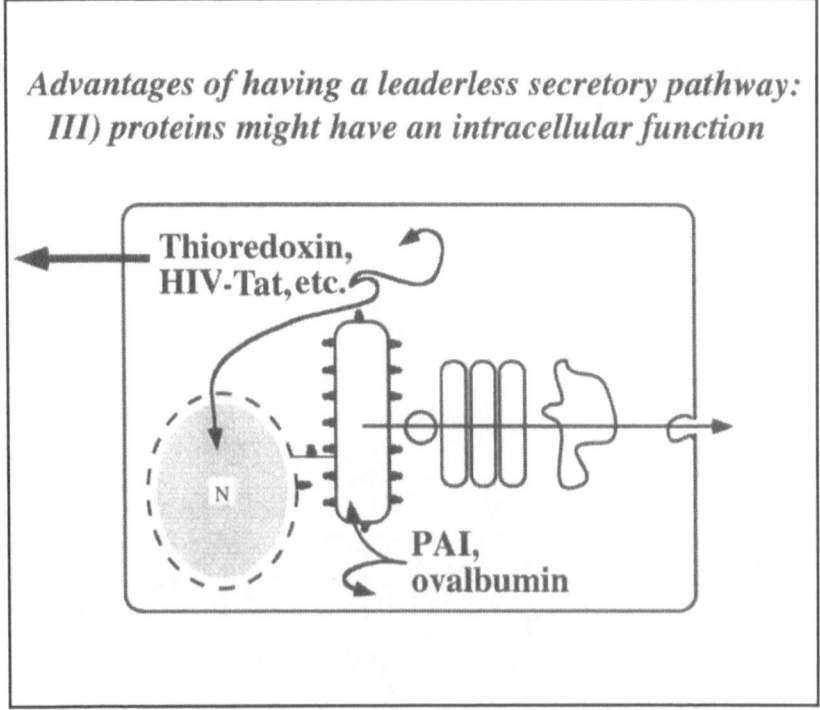

Advantages of having a leaderless secretory pathway: III) proteins might have an intracellular function

Fig. 3.6. The leaderless pathway allows the dual localization of the same polypeptide. Some leaderless proteins, such as thioredoxin or HIV-Tat have well defined cytosolic and/or nuclear functions in addition to their extracellular ones. The existence of a leaderless form of the interleukin-1 receptor antagonist, that is actually secreted by differentiated keratinocytes,[24] suggests that IL-1 may also have some intracellular function. Many mechanisms are exploited by cells to deliver the same polypeptide to two different compartments.[107] For instance, PAI and ovalbumin are found both in the cytosol and in the extracellular space, because they bear an internal signal sequence, less efficient than the typical N-terminal ones.

lumen; as many leaderless secretory proteins have free thiols, transit in the oxidizing milieu of the ER[90] might result in either retention and degradation or in a different folding, possibly causing inactivation of the protein (Fig. 3.8). For carbohydrate-binding proteins, such as lectins, the abundance of sugars moieties in the Golgi stacks may be detrimental for their transport.

Many leaderless secretory proteins are involved in the control of cell growth and differentiation. Two features of these proteins may compensate their slow secretion: (i) they have a high biological activity, minute amounts being sufficient in exerting a response in target cells; (ii) they act in a paracrine and/or autocrine way, on the producing cell or on nearest cells. Indeed, the slow and inefficient secretion, together with the binding of some of these proteins to low

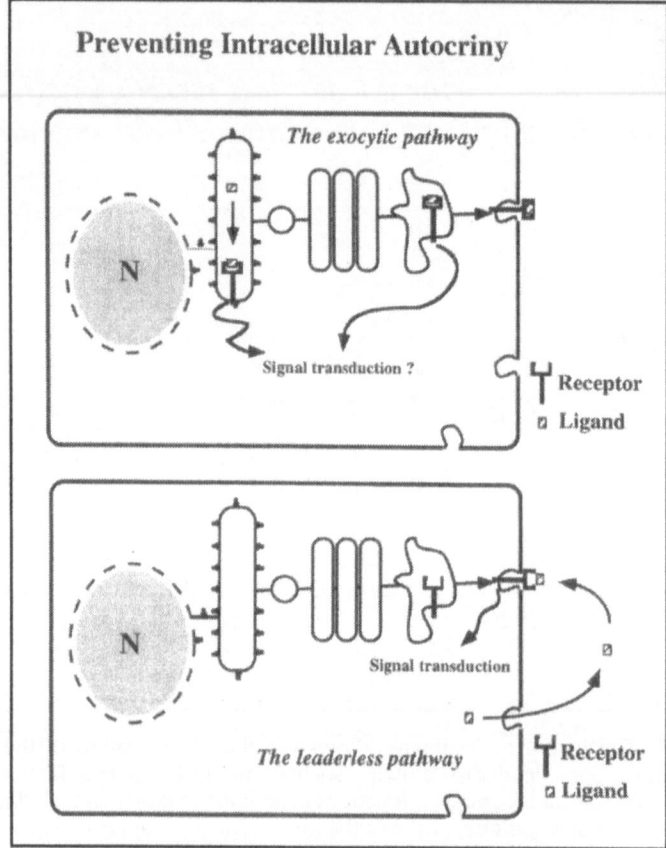

Fig. 3.7. Keeping the ligands apart from their receptors. In autocrine systems, the chances of interactions between a ligand and its receptor are bound to be much higher within vesicles of the exocytic pathway than on the plasmamembrane. Since there are many reports suggesting that signaling can be generated from receptors still in the ER,[108,109] compartmentalization may be essential to prevent intracellular autocriny. Similarly, the galactose-binding lectins that are secreted through the leaderless pathways, if entering the classical exocytotic route might interact in the Golgi with galactosylated membrane and soluble proteins.

affinity binding sites (such as heparan sulphates for FGFs and tat or carbohydrates for lectins) on cell surface or extracellular matrix, may serve to prevent these proteins from exerting a biological effect far from the site of production. Thus, the "short range" autocrine secretion of leaderless proteins might have a role in the local control of cell growth. Support of this hypothesis comes from the observation that loss of regulation of leaderless secretion in some cases plays

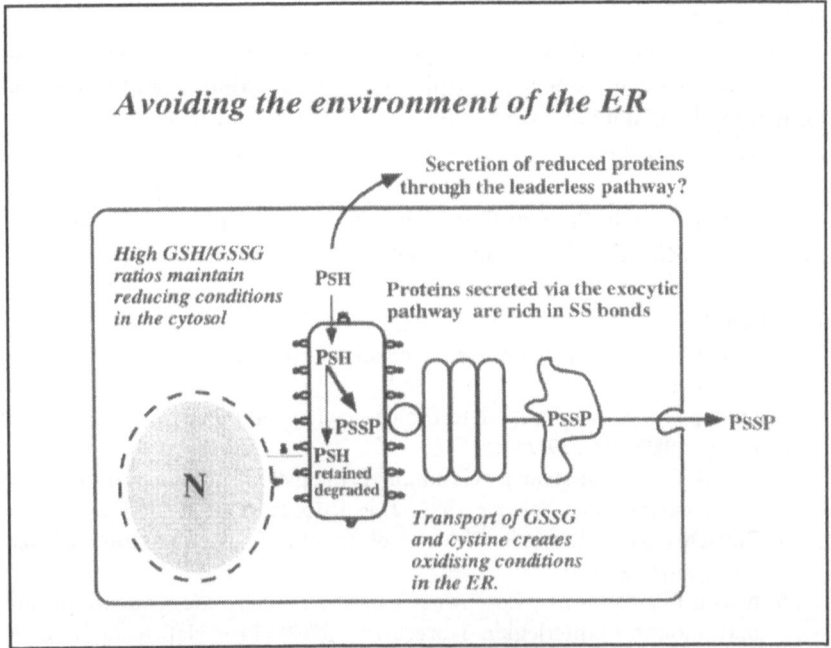

Fig. 3.8. Avoiding the oxidizing milieu of the exocytic compartment. One of the advantages offered by the leaderless secretory pathway(s) is avoiding the unique conditions of the ER lumen (e.g., high calcium concentrations).[110] The lumen of the ER is more oxidizing than the cytosol, mainly due to the existence of transporter molecules for oxidized glutathione (GSSG) and cystine.[90] As a result, proteins that are secreted through the exocytic pathway are rich in disulfide bridges. Since proteins with exposed thiols are retained and degraded in the ER,[111,112] the leaderless pathway may be essential to secrete proteins in the reduced state. Interestingly, many leaderless proteins (thioredoxin, FGF-1, HIV tat) are active only in the reduced state.

a role in altered cell proliferation. In this regard, either increased or decreased secretion of leaderless proteins has been reported to correlate with neoplastic cell growth: a secretory switch of FGF-2 accompanies the progression from fibroma to fibrosarcoma;[86] IL-1 is autocrinally secreted by acute myeloid leukemia cells; the inhibition of the autocrine circuit results in block of cell proliferation.[9] Conversely, hepatoma cells have a very low secretory rate of thioredoxin compared to normal liver cells; stimulation of thioredoxin secretion by reducing agents correlates with growth arrest.[52]

In the case of proteins which are also able to enter into neighbor cells such as Tat or FGF[38] perhaps the inefficiency of both outward and inward translocation is functional in generating local gradients during morphogenesis, wound healing or inflammation.

ACKNOWLEDGMENTS

We wish to thank Prof. C.E. Grossi and all our colleagues in Genoa and Milan, particularly N. Bonifaci, A. Cabibbo, S. Carelli, P. Morgavi and M. Pagani, for helpful discussions and suggestions, and S. Trinca for her outstanding secretarial assistance. Our work has been made possible through grants from the Associazione Italiana per la Ricerca sul Cancro (AIRC), Consiglio Nazionale delle Ricerche (CNR, Progetti Finalizzati ACRO, BTBS, IG), Fondazione San Raffaele and Ministero della Sanità (AIDS Special Project).

REFERENCES

1. Blobel G. Intracellular protein topogenesis. Proc Natl Acad Sci USA 1980; 77: 1496-1500.
2. Lingappa VR. Intracellular traffic of newly synthesized proteins. J Clin Invest 1989; 83: 739-751.
3. Schatz G, Dobberstein B. Common principles of protein translocation across membranes. Science 1996; 271: 1519-1526.
4. Bretscher MS, Munro S. Cholesterol and the Golgi apparatus. Science 1993; 261: 1280-81.
5. Auron PE, Webb AC, Rosenwasser LJ et al. Nucleotide sequence of human monocyte interleukin-1 precursor cDNA. Proc Natl Acad Sci USA 1984; 81: 7907-7911.
6. Abraham JA, Whang JL, Tumolo A et al. Human fibrolast growth factor: nucleotide sequence and genomic organization. EMBO J 1986; 5: 2523-2528.
7. Muesch A, Hartmann E, Rohde K et al. A novel pathway for secretory proteins? Trends Biochem Sci 1990; 15: 86-88.
8. Rubartelli A, Cozzolino F, Talio M, Sitia R. A novel secretory pathway for interleukin-1β, a protein lacking a signal sequence. EMBO J 1990; 9: 1503-1510.
9. Cozzolino F, Rubartelli A, Aldinucci D et al. Interleukin-1 as an autocrine growth factor for acute myeloid leukemia cells. Proc Natl Acad USA 1989; 86: 2369-2373.
10. Kuchler K. Unusual routes of protein secretion: the easy way out. Trends Cell Biol 1993; 3: 421-426
11. Dinarello CA. Interleukin-1α and Interleukin-1β antagonism. Blood 1991; 127: 119-127
12. Hazuda DJ, Lee JC, Young PR. The kinetics of interleukin-1 secretion from activated monocytes. Difference between interleukin-1α and interleukin-1β. J Biól Chem 1989; 17: 8473-8479.
13. Hogquist KA, Nett MA, Unanue ER et al. Interleukin-1 is processed and released during apoptosis. Proc Natl Acad Sci USA 1991; 88: 8485-8489.
14. Rubartelli A, Bajetto A, Allavena G et al. Post-translational regulation of interleukin-1β secretion. Cytokine 1993; 5: 117-124.

15. Perregaux D, Gabel CA. Interleukin–1β maturation and release in response to ATP and nigericin. Evidence that potassium depletion mediated by these agents is a necessary and common feature of their activity. J Biol Chem 1994; 269: 15195-15203.

16. Siders WA, Mizel SB. Interleukin-1 secretion. A multiple process that is regulated in a cell type-specific manner. J Biol Chem1995; 268: 22170-22174

17. Thornberry NA, Bull HG, Calaycay JR. A novel heterodimeric cysteine protease is required for interleukin-1β processing in monocytes. Nature 1992; 356: 768-774.

18. Cerretti DP, Kozlosky CJ, Mosley B. Molecular cloning of the interleukin-1β converting enzyme. Science 1992; 256: 97-100.

19. Miura M, Zhu H, Rotello R et al. Induction of apoptosis in fibroblasts by IL-1β-converting enzyme, a mammalian homolog of the *C. elegans* cell death gene ced-3. Cell 1993; 75: 653-660.

20. Ayala JM, Yamin TT, Egger LA et al. IL-1β-converting is present in monocytic cells as inactive precursor. J Immunol. 1994; 153: 2592-2599

21. Singer II, Scott S, Chin J et al. The interleukin-1β converting enzyme (ICE) is localized on the external cell surface membrane and in the cytoplasmic ground substance of human monocytes by immuno-electron microscopy. J Exp Med 1995; 182: 1447-1459.

22. Walev I, Reske K, Palmer M et al. Potassium-inhibited processing of IL-1β in human monocytes. EMBO J 1995; 14: 1607-1614.

23. Haskill S, Martin G, Van Le L et al. cDNA cloning of an intracellular form of the human interleukin-1 receptor antagonist associated with epithelium. Proc Natl Acad Sci USA 1991; 88: 3681-3685.

24. Corradi A, Franzi AT, Rubartelli A. Synthesis and secretion of interleukin-1α and interleukin-1 receptor antagonist during differentiation of cultured keratinocytes. Exp Cell Res 1995; 217: 355-362.

25. Burgess WH, Maciag T. The heparin-binding (fibroblast) growth factor family of proteins. Ann Rev Biochem 1989; 58: 575-560.

26. Basilico C, Moscatelli D. The FGF family of growth factors and oncogenes. Adv Cancer Res 1992; 59: 115-165.

27. Johnson DE, Williams LT. Structural and functional diversity in the FGF receptor multigene family. Adv Cancer Res 1993; 60: 1-41.

28. Jaye M, Howk R, Burgess W et al. Human endothelial cell growth factor: cloning, nucleotide sequence and chromosome localization. Science 1986; 233: 543-545.

29. Miyamoto M, Naruo KI, Seko C et al. Molecular cloning of a novel cytokine cDNA encoding the ninth member of the fibroblast growth factor family, which has a unique secretion property. Mol Cell Biol 1993; 13: 451-459.

30. Mignatti P, Morimoto T, Rifkin DB. Basic fibroblast growth factor realased by single, isolated cells stimulates their migration in an autocrine manner. Proc Natl Acad Sci USA 1991; 88: 11007-11011.

31. Mignatti P, Morimoto T, Rifkin DB. Basic fibroblast growth factor, a protein devoid of secretory signal sequence, is released by cells via a pathway independent of the endoplasmic reticulum-Golgi complex. J Cell Physiol 1992; 151: 81-93.

32. Jackson A, Tarantini F, Gamble S et al. The release of fibroblast growth factor-1 from NIH 3T3 cells in response to temperature involves the function of cysteine residues. J Biol Chem 1995; 270: 33-36.

33. Tarantini F, Gamble S, Friedman S et al. The cysteine residue responsible for the release of fibroblast growth factor-1 resides in a domain independent of the domain for phosphatidylserine binding. J Biol Chem 1995; 270: 29039-29042.

34. Imamura T, Tokita Y, Mitsui YJ. Identification of a heparin-binding growth factor-1 translocation sequence by deletion mutation analysis. J Biol Chem 1992; 267: 5676-5679.

35. Quarto N, Finger FP, Rifkin DB. The NH2-terminal extension of high molecular weight bFGF is a nuclear targeting signal. J Cell Physiol 1991; 147: 311-318.

36. Bouche G, Gas N, Prats M. Basic fibroblast growth factor enters the nucleolus and stimulates the transcription of ribosomal genes in ABAE cells undergoing G0/G1 transition. Proc Natl Acad Sci USA 1987; 84: 6770-6774.

37. Wiedlocha A, Falnes PO, Madshus IH et al. Dual mode of signal transduction by externally added acidic fibroblast growth factor. Cell 1994; 76: 1039-1051.

38. Rubartelli A, Sitia R. Entry of exogenous polypeptides into the nucleus of living cell: facts and speculations. Trends Cell Biol 1995; 5: 409-412.

39. Tabe L, Krieg P, Strachan R et al. Efficient expression of cloned complementary DNAs for secretory proteins after injection into *Xenopus* oocytes. J Mol Biol 1984; 180: 615-643.

40. Wiedmann B, Sakai H, Davis TA and Wiedmann M. A protein complex required for signal-sequence specific sorting and translocation. Nature 1994; 370: 434-440.

41. Li JJ, Huang YQ, Moscatelli D. Expression of fibroblast growth factors and their receptors in acquired immunodeficiency syndrome-associated Kaposi sarcoma tissue and derived cells. Cancer 1993; 72: 2253-2259.

42. Rogelj S, Weinberg RA, Fanning P, Klagsbrun M. b-FGF fused to a signal peptide transforms cells. Nature 1988; 331: 173-175.

43. Holmgren A. Thioredoxin. Ann Rev Biochem 1985; 54: 237-271.

44. Holmgren A. Thioredoxin and glutaredoxin systems. J Biol Chem 1989; 264: 13963-13966.

45. Wollman EE, D'Auriol L, Rimsky L et al. Cloning and expression of a cDNA for human thioredoxin. J Biol Chem 1988; 263: 15506-15512.

46. Tagaya Y, Maeda Y, Mitsui A et al. ATL-derived factor (ADF), an IL-2 receptor/Tac-inducer homologous to thioredoxin: possible involvement of dithiol-reduction in the IL-2 receptor induction. EMBO J 1989; 8: 757-764.

47. Wakasugi N, Tagaya Y, Wakasugi H et al. ADF/thioredoxin produced by both HTLV-1 and EBV transformed lymphocytes acts as an autocrine growth factor and synergizes with IL-1 and IL-2. Proc Natl Acad Sci USA 1990; 87: 8282-8286.
48. Silverstein DS, Ali MH, Baker SL et al. Human eosinophil cytotoxicity-enhancing factor purification, physical characteristics, and partial aminoacid sequence of an active polypeptide. J Immunol 1989; 143: 979-983.
49. Carlsson M, Totterman TH, Rosen A et al. Interleukin 2 and a cell hybridoma (MP6) derived factor act synergistically to induce proliferation and differentiation of human B-chronic lymphocytic leukemia cells. Leukemia 1989; 3: 593-601.
50. Rubartelli A, Bajetto A, Allavena G et al. Secretion of thioredoxin by normal and neoplastic cells through a leaderless secretory pathway. J Biol Chem1992; 267: 24161-24164.
51. Ericson ML, Horling J, Wendel HV et al. Secretion of thiredoxin after in vitro activation of human B cells. Lymphokine-Cytokine Res 1992; 11: 201-207.
52. Rubartelli A, Bonifaci N, Sitia R. High rates of thioredoxin secretion correlate with growth arrest in hepatoma cells. Cancer Res 1995; 55: 675-680.
53. Martin H, Dean M. Identification of a thioredoxin related protein associated with plasma membrane. Biochem Biophys Res Commun 1991; 175: 123-128.
54. Barondes SH. Soluble lectins: a new class of extracellular proteins. Science 1984; 223: 1259-1264.
55. Cherayil BJ, Chaitovitz S, Wong C et al. Molecular cloning of a human macrophage lectin specific for galactose. Proc Natl Acad Sci USA 1990; 87: 7324-7328.
56. Rosenberg IM, Lyer R, Cherayil BJ et al. Structure of the murine *Mac2* gene. J Biol Chem 1993; 268: 12393-12400.
57. Raz A, Meromsky L, Zvibel I et al. Transformation-related changes in the expression of endogenous cell lectins. Int J Cancer 1987; 388: 353-360.
58. Wells V, Mallucci L. Identification of an autocrine negative growth factor: mouse β-galactoside-binding protein is a cytostatic factor and cell growth regulator. Cell 1991; 64: 91-97.
59. Barondes SH, Haywood-Reid PL. Externalization of an endogenous chicken muscle lectin with in vivo development. J Cell Biol 1981; 91: 568-572.
60. Cooper DN, Barondes SH. Evidence for export of a muscle lectin from cytosol to extracellular matrix and for a novel secretory mechanism. J Cell Biol 1990; 110: 1681-1691.
61. Johnstone RM, Adam M, Hammond JR et al. Vesicle formation during reticulocyte maturation: association of plasmamembrane activities with released vesicles (exosomes). J Biol Chem 1987; 262: 9412-9420.

62. Hale JE, Wuthier RE. The mechanism of matrix vesicles formation. J Biol Chem 1987; 262: 1916-1925.
63. Lindstedt R, Apodaca G, Barondes SH et al. Apical secretion of a cytosolic protein by Madin-Darby canine kidney cells. Evidence for polarized release of an endogenous lectin by a nonclassical secretory pathway. J Biol Chem 1993; 268: 11750-11757.
64. Sato S, Burdett I, Hughes RC. Secretion of the baby hamster kidney 30 kDa galactose binding lectin from polarized and nonpolarized cells: a pathway independent of the endoplasmic reticulum-Golgi complex. Exp Cell Res 1993; 207: 8-18
65. Jones KA. Tat and HIV-1 promoter. Curr Opin Cell Biol 1993; 5: 461-468.
66. Westerndorp MO, Li-Weber M, Frank RW et al. Human immunodeficiency virus type 1 Tat upregulates interleukin-2 secretion in activated T cells. J Virol 1994; 68: 4177-4185.
67. Ensoli B, Barillari G, Salahuddin SZ et al. Tat protein of HIV-1 stimulates growth of cells derived from Kaposi's sarcoma lesions of AIDS patients. Nature 1990; 345: 84-86.
68. Westendorp MO, Frank R, Ochsenbauer C et al. Sensitization of T cells to CD95-mediated apoptosis by HIV-1 Tat and gp120. Nature 1995; 375: 497-500.
69. Albini A, Benelli R, Presta M et al. HIV-tat protein is a heparin-binding angiogenic growth factor. Oncogene 1996; 12: 289-297.
70. Frankel AD, Pabo CO. Cellular uptake of the tat protein from human immunodeficiency virus. Cell 1988; 55: 1189-1193.
71. Mann DA, Frankel AD. Endocytosis and targeting of exogenous HIV-1 Tat protein. EMBO J 1991; 10: 1733-1739.
72. Dice JF. Peptide sequences that target cytosolic proteins for lysosomal proteolysis. Trends Biochem Sci 1990; 15: 305-309.
73. Terlecky SR, Dice JF. Polypeptide import and degradation by isolated lysosomes. J Biol Chem 1993; 268: 23490-23495.
74. Cuervo AM, Dice JF. A receptor for the selective uptake and degradation of proteins by lysosomes. Science 1996; 273: 501-503.
75. Hendrick J, Hartl FU. Molecular chaperone function of heat shock proteins. Ann Rev Biochem 1993; 62: 349-384.
76. Randall L, Hardy S. Correlation of competence for export with lack of tertiary structure in the mature species: a study in vivo of maltose binding protein in _E. coli._ Cell 1986; 46: 921-928.
77. Eilers M, Schatz G. Binding of a specific ligand inhibits import of a purified precursor protein into mitochondria. Nature 1986; 322: 228-232.
78. Deshaies RJ, Koch BD, Werner-Washburne M et al. A family of stress proteins facilitates translocation of secretory and mitochondrial precursor proteins. Nature 1988; 322: 800-810.
79. McNew JA, Goodman JM. An oligomeric protein is imported into peroximoses in vivo. J Cell Biol 1994; 127: 1245-1257.
80. America T, Hageman J, Guère A et al. Methotrexate does not block

import of a DHFR fusion protein into chloroplasts. Plant Mol Biol 1994; 24: 283-294.

81. Hazuda DJ, Strickler J, Simon P, Young PR. Structure-function mapping of interleukin-1 precursors. Cleavage leads to a conformational change in the mature protein. J Biol Chem 1991; 266: 7081-7086.

82. Flower RJ, Rothwell NJ. Lipocortin 1: cellular mechanisms and clinical relevance. Trends Pharmacol Sci 1994; 15: 71-76.

83. Silver PA. How proteins enter the nucleus. Cell 1991; 64: 489-497.

84. Ho KC, Quarmby VE, French FS, Wilson EM. Molecular cloning of rat prostate transglutaminase cDNA: the major androgen-regulated protein DP1 of rat dorsal prostate and coagulating gland. J Biol Chem 1992; 267: 12660-12667.

85. Seitz J, Keppler C, Rausch U, Aumuller G. Immunohistochemistry of secretory transglutaminase from rodent prostate. Histochemistry 1990; 93: 525-530.

86. Kandel J, Bossy-Wetzel E, Radvanyi F et al. Neovascularization is associated with a switch to the export of bFGF in the multistep development of fibrosarcoma. Cell 1991; 66: 1095-1104.

87. Bohmer FD, Kraft R, Otto A et al. Identification of a polypeptide growth inhibitor from bovine mammary gland. J Biol Chem 1987; 262: 1537-1543.

88. Cozzolino F, Torcia M, Aldinucci D et al. Interleukin-1 as an autocrine regulator of human endothelial cell growth. Proc Natl Acad Sci USA 1990; 87: 6487-6491.

89. Deiss LP, Kimchi A. A genetic tool used to identify thioredoxin as a mediator of a growth inhibitory signal. Science 1991; 252: 117-120.

90. Hwang C, Sinskey AJ and Lodish HF. Oxidized redox state of glutathione in the endoplasmic reticulum. Science 1992; 257: 1496-1502.

91. Lin L-F H, Mismer D, Lilie JD et al. Purification, cloning and expression of ciliary neurotrophic factor (CNTF). Nature 1989; 246: 1023-1025.

92. Ishikawa F, Miyazono K, Hellman U et al. Identification of angiogenic activity and the cloning and expression of platelet-derived endothelial cell growth factor. Nature 1989; 388: 557-562.

93. Usuki K, Heldin NE, Miyazono K et al. Production of platelet-derived endothelial cell growth factor by normal and transformed human cells in culture. Proc Natl Acad Sci USA 1989; 86: 7427-7431.

94. Kao J, Houck K, Fan Y et al. Characterization of a novel tumor-derived cytokine. J Biol Chem 1994; 269: 25106-25119.

95. Christmas P, Callaway J, Fallon J et al. Selective secretion of annexin 1, a protein without a signal sequence, by the human prostate gland. J Biol Chem 1991; 266: 2499-2507.

96. Solito E, Raugei G, Melli M et al. Dexamethasone induces the expression of the mRNA of lipocortin 1 and 2 and the release of lipocortin 1 and 5 in differentiated but not in undifferentiated U-937 cells. FEBS Lett 1991; 291: 238-244.

97. Goodall GJ, Dominguez F, Horecker BL. Molecular cloning of cDNA for human prothymosin α. Proc Natl Acad Sci USA 1986; 83: 8926-8928.

 98. Clinton M, Frangou-Lazaridis M, Pannerselvam C et al. The sequence of human parathymosin deduced from a cloned human kidney cDNA. Biochem Biophys Res Commun 1989; 158: 855-862.
 99. Grundmann U, Amann E, Zettlmeissl G et al. Characterization of cDNA coding for human factor XIIIa. Proc Natl Acad Sci USA 1986; 83: 8024-8028.
100. Gentile V, Saydak M, Chiocca EA et al. Isolation and characterization of cDNA clones to mouse macrophage and human endothelial cell tissue transglutaminases. J Biol Chem 1991; 266: 478-483.
101. Aeschlimann D, Paulsson M. Transglutaminases: protein cross-linking enzymes in tissue and body fluids. Thromb Haemos 1994; 71:402-415.
102. Melloni E, Sparatore B, Patrone M, Pessino A, Passalacqua M, Pontremoli S. Extracellular release of the "differentiation enhancing factor", a HMG1 protein type, is an early step in murine erythroleukemia cell differentiation. FEBS Lett 1995; 368: 466-470.
103. Sloan IS, Horowitz PM, Chirgwin JM. Rapid secretion by a non classical pathway of overexpressed mammalian mitochondrial rhodanese. J Biol Chem 1994; 269: 27625-27630.
104. Biocca S, Cattaneo A. Intracellular immunization: antibody targeting to subcellular compartments. Trends Cell Biol 1995; 5: 248-252.
105. Simon SM, Blobel G. Signal peptides open protein-conducting channels in *E. coli*. Cell 1992; 69: 677-684.
106. Ooi CH, Weiss J. Bidirectional movement of a nascent polypeptide across microsomal membranes reveals requirements for vectorial translocation of proteins. Cell 1992; 71: 87-96.
107. Danpure CJ. How can the product of a single gene be localized to more than one intracellular compartment. Trends Cell Biol 1995; 5: 230-238.
108. Dunbar CE, Browder TM, Abrams JS, Nienhuis AW. COOH-terminal modified interleukin 3 is retained intracellularly and stimulates autocrine growth. Science 1989; 245: 1493-1496.
109. Bejcek BE, Li DJ, Deuel TF. Transformation by v-sis occurs by an internal autoactivation mechanism. Science 1989; 245: 1496-1499.
110. Montero M, Brini M, Marsault R et al. Monitoring dynamic changes in free Ca2+ concentration in the endoplasmic reticulum of intact cells. EMBO J 1995; 14: 5467-5475.
111. Sitia R, Neuberger MS, Alberini C et al. Developmental regulation of IgM secretion: the role of the carboxy-terminal cysteine. Cell 1990; 60; 781-790.
112. Reddy P, Sparvoli A, Fagioli C et al. Formation of reversible disulfide bonds with the protein matrix of the endoplasmic reticulum correlates with the retention of unassembled Ig-light chains. EMBO J 1996; 15: 2077-2085.

THE TRANSPORTERS ASSOCIATED WITH ANTIGEN PROCESSING (TAP)

Robert Tampé, Stefanie Urlinger, Kurt Pawlitschko and Stephan Uebel

I. INTRODUCTION

Class I molecules of the major histocompatibility complex (MHC) present peptides derived from endogenous proteins at the cell surface. During viral infection or malignant transformation a different set of peptides is displayed by MHC class I molecules. These antigen-loaded class I complexes are recognized by cytotoxic T cells via the T cell receptors as nonself, thus leading to the destruction of the abnormal cell.

Every individual expresses only three to six different class I alleles. Each of them must be able to bind a variety of different peptides to ensure efficient elimination of infected or transformed cells. MHC class I molecules consist of two subunits: the membrane-anchored heavy (H) chain, which is highly polymorphic, and a noncovalently associated, soluble β_2-microglobulin (β_2m). The α_1 and α_2 domains of the heavy chain form the peptide binding groove of the molecule, whereas β_2m binds to the α_3 domain. X-ray crystallography of different class I MHC molecules together with the corresponding peptide ligands gave insight into the fine structure of the peptide binding groove.[1-3] The α_1 and α_2 domains form a β-sheet platform, which is flanked at two opposite sides by two long α-helices. The ends of the binding groove are closed and bury the N- and C-termini of bound peptides through hydrogen bond interactions via conserved amino acids. This normally limits the length of the peptide ligands to 8-10 residues. However, even longer peptides were found to associate with class I molecules. But in these cases either the central parts of the peptides were bulging out,[4] or one end of the peptide was extending out of the groove.[5]

How are these antigenic peptides generated? Assembly of class I molecules occurs in the endoplasmic reticulum (ER), but the majority of antigenic peptides presented at the cell surface is derived from cytosolic or nuclear proteins

Unusual Secretory Pathways: From Bacteria to Man, edited by Karl Kuchler.
© 1997 R.G. Landes Company.

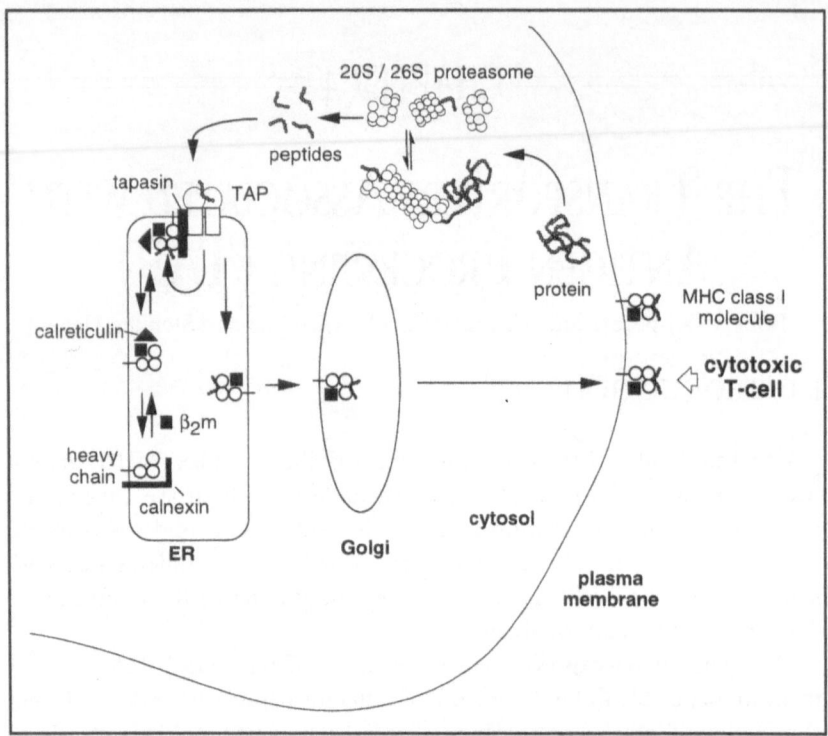

Fig. 4.1. Antigen presentation via MHC class I molecules.

(Fig. 4.1). The 26S proteasome and its catalytic 20S core are known to be responsible for the major proteolytic activities in the cytosol and nucleus, and the role of this multicatalytic proteinase in antigen processing has been extensively investigated. First evidence for the involvement of the proteasome complex in antigen processing was the discovery of two genes within the MHC II locus, namely *lmp2* and *lmp7* (low molecular mass polypeptide), which were found to encode proteasomal subunits.[6-8] The expression of these proteins is induced by γ-interferon and, once present, Lmp2 and Lmp7 assemble to form the 20S proteasome complex by substitution of constitutive subunits.[9-11] The in vitro characterization of Lmp-containing 20S-complexes[12-14] and in vivo experiments with Lmp2 or Lmp7 knockout mice[15,16] gave inconsistent results. The effects observed are subtle and the functional role of these subunits in antigen processing remains open. Furthermore, Rock and coworkers showed that inhibitors of the 20S proteasome, as well as mutations in the ubiquitination pathway, impair the intracellular assembly of class I molecules and the presentation of peptide epitopes in vivo,[17,18] due to inhibition of intracellular protein degradation.

As soon as peptides of the appropriate length are generated in the cytosol, delivery to the site of chaperone-assisted formation of the trimeric MHC-peptide complex, the ER lumen, must take place. First evidence for the existence of a peptide transporter came from mutant cell lines, showing decreased MHC class I surface expression.[19-21] The gene defects were located to the MHC class II region and led to the cloning and characterization of a novel group of MHC genes, which showed high homology to other known transport proteins, and were finally named *tap1* and *tap2* (Transporter associated with antigen processing).[22-25] These two genes were found in direct proximity to the genes encoding the proteasomal subunits Lmp2 and Lmp7 and, like them, were also shown to be inducible by γ-interferon. Transfection of the missing *tap* genes into the defect cell lines resulted in restored MHC class I antigen presentation[26-28] and thus gave the first evidence that TAP is directly involved in antigen processing. Further studies showed that *tap1* and TAP2 form stable heterodimeric complexes,[29] and localized the proteins to the ER and cis-Golgi compartments.[30] Early experiments directly investigating peptide transport across microsomal membranes raised doubt about the role of TAP, since it was reported to be ATP-and TAP-independent.[31,32] However, later studies dealing either with permeabilized cells[33,34] or isolated microsomes[35] from mammalian cell lines gave strong evidence that peptide transport was strictly dependent on both ATP and TAP. Furthermore, heterologous expression of TAP in insect cells demonstrated that this protein complex can mediate ATP-dependent peptide transport across microsomal membranes without the help of any additional factors derived from a highly developed immune system.[36]

Although the *tap* genes are polymorphic with several *tap1* and *tap2* alleles found in humans,[37-39] a functional polymorphism could only be observed in rat. Here, different *tap2* alleles are associated with an altered spectrum of peptides bound to the MHC molecules,[40] as well as with the accumulation of different sets of peptides in isolated microsomes[41] and different peptide binding patterns to TAP.[42]

II. STRUCTURAL ORGANIZATION OF TAP

The TAP complex belongs to the ATP-binding cassette (ABC) transporter superfamily of proteins, members of which typically consist of four domains: two hydrophobic transmembrane regions, where each is composed of predicted five to eight helices, and two hydrophilic parts facing the cytosol.[43] The designation of this protein family is derived from the highly conserved domains, where the Walker motifs A and B form the site for ATP-binding and hydrolysis, therefore called nucleotide binding domains (NBD).[44] One characteristic feature of the ABC transporter is the C-loop, a conserved region of about six to eight amino acids located several residues upstream of the Walker B motif, which is supposed to be involved in energy coupling to the transmembrane domains (TMO). An additional sequence motif found in almost all ABC transporters is localized in a the last cytoplasmic loop of the membrane spanning domain.[45] There is evidence that this "EAA-like" motif of about twenty amino acids interacts with the nucleotide binding fold, especially via the C-loop, because mutations in this region change or abolish transport activity just in the same way as in NBD mutants.[46,47] All four domains can be encoded by one gene or they can be linked in different combinations. The transporter associated with antigen processing is a heterodimer (TAP1·TAP2) with each half consisting of one transmembrane and one nucleotide binding domain.

To inspect the relationship of the TAP proteins to other ABC transporters we performed two independent phylogenetic analyses, one with the nucleotide binding domain (Fig. 4.2a) and the other with the transmembrane domain (Fig. 4.2b). For clearness, we focused on eukaryotic ABC transporters (except the bacterial hemolysin transporter HlyB). Furthermore, only one representative of each species or subfamily was given. Two members of the P-glycoprotein subfamily (MDR1 and MDR3) were included to illustrate the inner relation-

Fig. 4.2. (opposite page) Dendrograms of TAP and related ABC transporters. A BLAST homology search of the National Center for Biotechnology Information (NCBI) collection of protein sequence databases was performed based on the protein sequences of human TAP1 and TAP2, human transporter associated with antigen processing, TAP1 and TAP2;[114] human multidrug resistance protein, MDR1 and MDR3;[115,116] human cystic fibrosis transmembrane conductance regulator, CFTR;[117] human adrenoleucodystrophy protein, ALDp;[118] human peroxisomal membrane protein, Pmp70;[119] human sulfonylurea receptor, SUR;[120] yeast a-factor pheromone transporter, Ste6;[121,122] the MDL1 and MDL2 gene products from S. cerevisiae;[123] mitochondrial transporter from S. cerevisiae, Atm1;[124] heavy metal tolerance protein from S. pombe, Hmt1;[125] hemolysin A transporter from E. coli, HlyB.[126] All sequences were obtained from the SWISSPROT database for protein sequences. For proteins containing two NBDs and two TMDs each domain was analyzed separately, with (N) and (C) signifying the N- and C-terminal halves, respectively. Taxonomic relationship was computed using the EClustalW program from the EGCG extensions to the University of Wisconsin Genetic Computer Group (GCG) package (version 8.1).[127] The resultant matrix was displayed using the DRAWTREE program from the PHYLIP phylogeny inference package (version 3.5c).[128] (A) This dendrogram shows the relationship of the proteins with regard to their nucleotide binding domains. Here sequences extending from 50 amino acids upstream of the Walker A site to 50 amino acids downstream of the Walker B site were compared. (B) Dendrogram concerning the transmembrane domains. The membrane spanning regions from helices 1 to 6 were defined using the TopPred II program (version 1.3).[129] These portions of the protein sequences were aligned with the MACAW program (version 2.0.3.), provided by the NCBI and a defined block containing 221 aligned amino acids was used to perform the taxonomic computation.

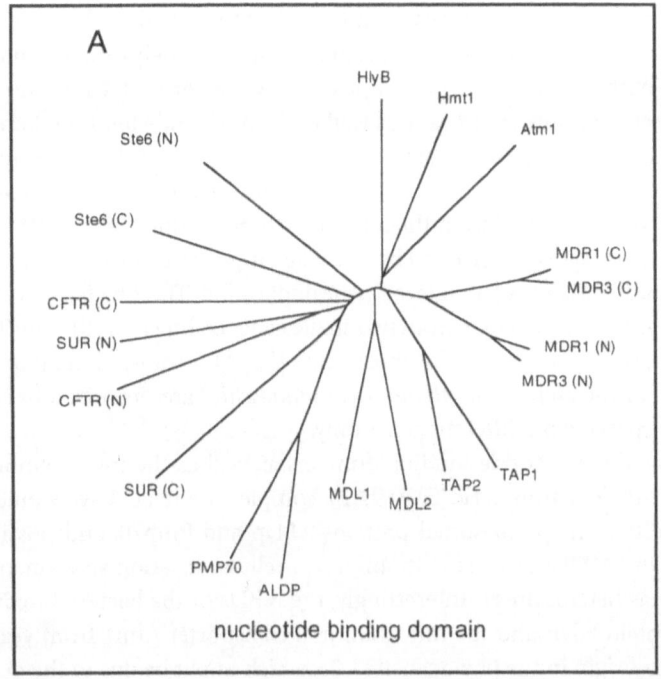

nucleotide binding domain

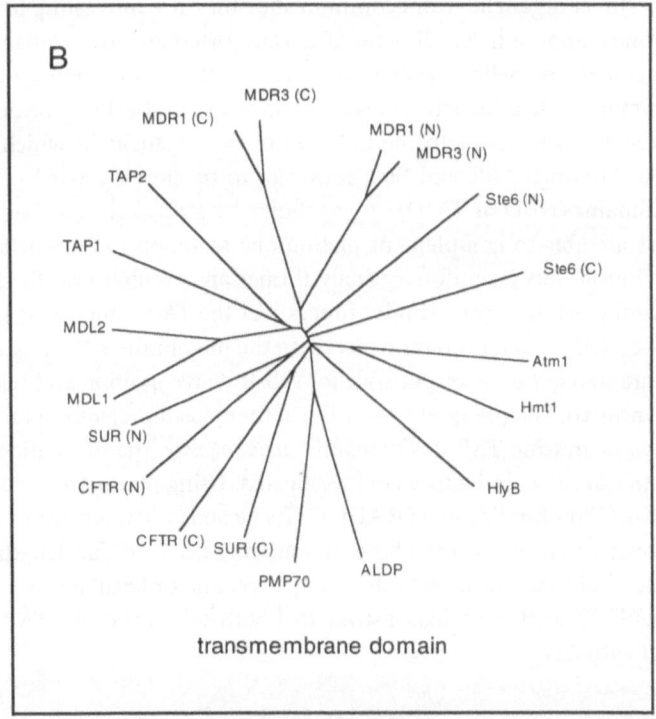

transmembrane domain

ship of these proteins. Interestingly, the N- and C-terminal domains of ABC transporters encoded by one gene product are not as closely related as one would expect. Within one subfamily composed of MDR- or CFTR-related proteins, the N-terminal transmembrane as well as the nucleotide binding domains are more related to each other than to their counterparts of the C-terminal half. This observation is in agreement with taxonomic studies performed only with a highly conserved region in the nucleotide binding domain.[48-50] We conclude that ABC transporters encoded by one gene originate from a fusion of two related genes[51] rather than an internal gene duplication. These half-size ABC transporters appear to originate from two single genes coding for NBD and TMD. In line with this hypothesis is the observation that ABC transporters found in the genome of the archeon *Methanococcus jannaschii*[52] are organized in NBDs or TMDs separated on different genes only.

Since the nucleotide binding domains as well as the transmembrane regions of the heterodimeric TAP1·TAP2 complex are related, we would like to propose that the peroxisomal proteins ALDp and Pmp70, and possibly also MDL1 and MDL2 of yeast with unknown cellular location and function may operate as heterodimers. Interestingly, the NBDs of the bacterial toxin transport protein HlyB and the mitochondrial transporter Atm1 from yeast share the same origin in the phylogenetic tree, which might be due to the descent of the subcellular organelles from common ancestors. It is interesting to note that eukaryotic homo- or heterodimeric ABC transporters are predominantly found in subcellular organelles. Testing the TMDs of the two proteins, a divergent development of these structures becomes obvious, maybe due to an adaptation to different substrates or to the function of the organelle in which they are inserted. Although Mdr and Ste6 seem not to be closely related concerning both domains (NBD or TMD), it was shown by several groups that Mdr-like proteins are able to complement pheromone secretion in Ste6-deleted yeast cells.[53-55] From this evolutionary analysis one can speculate that the Mdl proteins from yeast perform a similar function as the TAP complex, as they seem to be the most related proteins concerning the membrane-spanning domains, which are thought to be responsible for substrate recognition and selection.

By means of heterologous expression of the cytosolic regions of human TAP in *E. coli* or murine TAP in *Drosophila melanogaster*, the nucleotide binding properties of these domains were investigated. Using photo-crosslinking with 8-azido-ATP, the binding of ATP, ADP, GTP, ITP and CTP was demonstrated.[56,57] These results were verified by photo-crosslinking studies of full-length proteins expressed in human or insect cells using a vaccinia or baculovirus expression system.[58,59] These studies demonstrate that both subunits of the TAP complex interact with ATP.

A crucial question with regard to the transport activity of TAP was whether binding of nucleotides is both necessary and sufficient for substrate translocation. In microsomes or permeabilized cells peptide translocation across the ER membrane was only detected in the presence of ATP, but not of nonhydrolyzable ATP analogs.[33-35] Peptides can bind to the transporter in the absence of nucleotides, but are not translocated.[60,61] A first functional mutation in the NBD of the human MHC-encoded transporter was found in a small cell lung carcinoma.[62] This single arginine to glutamine exchange at position 659 of TAP1 results in the loss of MHC expression on the cell surface. As this amino acid substitution is located directly between the C-loop and the Walker B motif, it is supposed to interfere with ATP-binding, ATP-hydrolysis or energy coupling to the transmembrane domain.

The transmembrane region is not as conserved as the nucleotide binding domain, where a sequence identity of about 30 to 50% is found between different ABC transporters. Prediction of the membrane topology from the sequences of several members of the ABC family revealed the variability in the number of transmembrane helices (TM): The hemolysin transporter HlyB is predicted to have eight transmembrane helices[63] and the maltose transporter subunits MalG and MalF are supposed to contain six and eight transmembrane helices, respectively.[64,65] Some examples of the eukaryotic members of this family seem to share a constant number of six transmembrane helices in each of the TMDs.

The membrane topology has been investigated by a combination of genetic and biochemical approaches: The membrane organization of CFTR has been mapped by the insertion of N-glycosylation consensus sequences (NXS/T) in predicted loops of the TMDs and subsequent examination of their accessibility to the glycosylation machinery in the ER.[66] Geller and coworkers have recently tested gene-fusion assays on Ste6 in both a heterologous prokaryotic and the native eukaryotic expression system (*E. coli* and *S. cerevisiae*, respectively).[67] As reporter molecules, alkaline phosphatase (AP) and invertase (Inv) were used, since both are accessible to substrates only when oriented extracellularly. In summary, these studies confirmed the prediction of a two-times-six helices motif spanning the membrane with the N- and C-termini facing the cytosol. Alternatively, membrane topology of MDR was addressed by insertion of antigenic peptide epitopes,[68,69] also verifying this pattern of helix insertion. However, the model of six transmembrane helices in each half of an ABC transporter seems not to be valid for all eukaryotic ABC proteins. By insertion of N-glycosylation targeting sequences into hamster P-glycoprotein, Ling and colleagues found a membrane topology, which seems to have two different orientations. The first orientation, as expected, has all six helices spanning the membrane, but in the second, only four helices were observed; predicted TM3 was found to form a larger intracellular loop and TM5 was located extracellularly.[70,71]

Nevertheless, the disturbance of the natural structure of a protein by all these insertions can be detrimental to functionality. A very elegant but laborious method to map membrane topology is to create a cysteine-less mutant. Then, after mutating single amino acids in predicted loops into a cysteine, it can be labeled alternatively by membrane permeable or impermeable SH-specific reagents. This approach is less invasive than insertion studies and every single amino acid can be studied. The reliability of this method was shown on Mdr[72,73] and also on a non-ABC transporter, the lactose permease from *E. coli.*[74,75] The transmembrane topology of the TAP complex has not yet been addressed experimentally. Sequence alignments with ABC transporters and the computer prediction of transmembrane helices support a model of the membrane topology of TAP:TAP1 of man and rat where they seem to span the membrane 8 to 10 times but only the last six helices are homologous to the TMDs of the shorter mouse TAP and other ABC transporters. The structure and function of this highly diverse N-terminal region of 170 to 180 amino acids remain unclear.

As mentioned above, sequence polymorphism was found in the *tap* genes of different organisms, but only in rat cells an altered antigenicity was found. The cim (class I modifier) polymorphism in rat is based on four *tap2* alleles designated a and l in cim[a], and c and u in cim[b]. MHC haplotypes with cim[b] type transporters are permissive for peptides containing hydrophobic C-terminal residues, whereas cim[a] type molecules show a more broad specificity including hydrophobic and basic C-terminal residues of the peptide substrate.[41,76] About 25 substitutions were found in the amino acid sequence correlating with cim type, with only two residues situated in the C-terminal half of the protein corresponding to the nucleotide binding domain.[37] By constructing several rat TAP2[a]-TAP2[u] chimera, four positions (217/218 and 374/380) located within the last two cytoplasmic loops of rat TAP2 were identified that are involved in peptide selection.[77] In contrast, in human TAP the polymorphic sites are found either within predicted transmembrane helices or the NBDs. This might be the reason that they have no influence on peptide binding.

The contribution of both TAP subunits to substrate binding was revealed by photo-activatable crosslinker peptides.[78] Using photo crosslinking and enzymatic digestion, a large region of human TAP1 (position 362 to 487) including the last cytoplasmic loop and the last two predicted α-helices was shown to be involved in peptide binding (Fig. 4.3).[79] In addition, by transferring functional polymorphic residues of rat TAP2, a A374D mutation of human TAP2 was identified that alters peptide selectivity of the TAP complex.[80] These observations are in agreement with substrate binding sites identified within other ABC transporters, as for example in MDR1, where substrate crosslinking and point mutations altering substrate specificity were found within similar regions.[81,82]

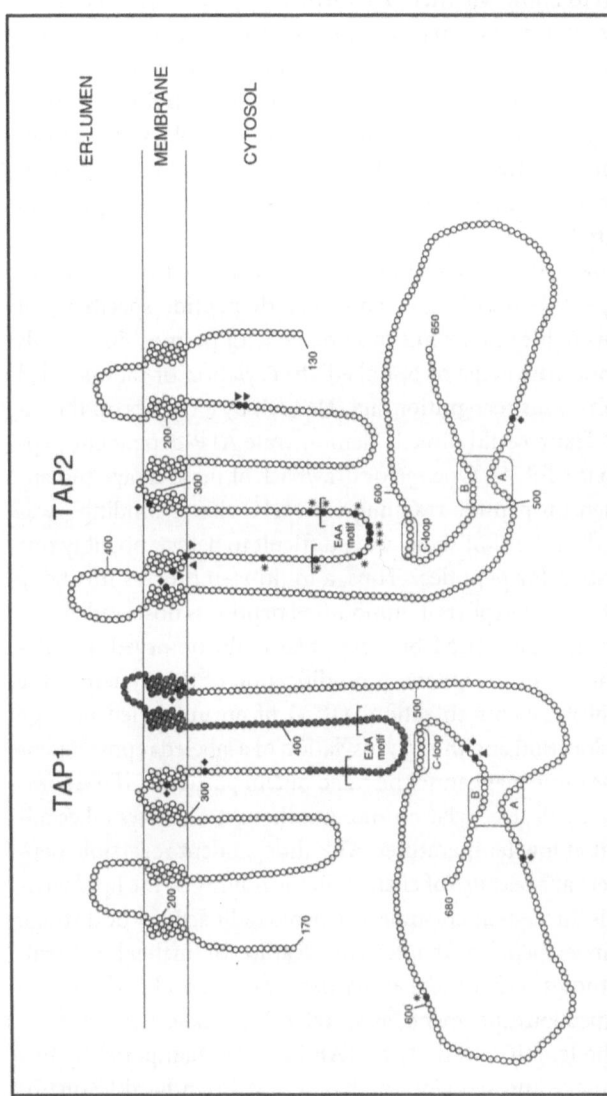

Fig. 4.3. Structural organization of TAP1 and TAP2. Topological model of human TAP1 and TAP2 based on the prediction of amphiphilic transmembrane domains (TopPred II, version 1.3) illustrating only six transmembrane helices, which are in homology to other ABC transporters, and the nucleotide binding domains. Each circle represents an amino acid residue, with filled-in circles marking several, but not all point mutations altering transport function that were derived from homologous positions in other ABC proteins: (▲) human TAP[79,80] (▼) rat TAP[77], (◆) human P-glycoprotein (summarized in[130], (✱) human CFTR.[47,131] The ATP-binding sites are circled with A and B indicating the Walker motifs. Gray circles show the photo-affinity labeled region of human TAP1 (see text for details).[79]

III. SUBSTRATE SELECTIVITY OF TAP

When it became clear that the TAP complex is the peptide translocator in the ER, interest soon focused on substrate specificity of TAP. In particular, immunologists were keen to know whether TAP further restricts the pool of peptides available for presentation on MHC class I-molecules, a question of certain interest for the generation of vaccines designed to stimulate cytotoxic T cell response. Furthermore, the mechanism of peptide selection for translocation by TAP is also of general biochemical interest since it displays high affinities known from highly selective receptor-ligand interactions but is different from the key-and-lock type of interaction in that a broad spectrum of peptides is effectively translocated.

Several assay systems have been developed so far to confirm the function of the TAP complex as peptide translocator and to study peptide specificity of TAP. A major breakthrough came from experiments with permeabilized cells or microsomes that took advantage of N-linked glycosylation of radiolabeled peptides carrying a consensus recognition site (NXS/T) by enzymes on the lumenal side of the ER. These could directly demonstrate ATP-dependent peptide translocation into the ER.[34,36] One major drawback of these assays and another system that relied on peptide trapping via MHC class I-binding[35] was that peptides without the retention signal were difficult to detect, possibly due to an active export system for peptides. Thus, a multi-step process involving peptide import, retention and export of unmodified peptide is observed and at least for assays relying on MHC class I-binding a bias in the observed specificity is introduced by the retention system. A modification of this system is the determination of inhibitory concentrations (IC_{50}) of an unlabeled peptide needed to reduce translocation and thus glycosylation of a labeled reporter peptide by 50%,[33,34] but the same bias cannot be ruled out, in particular if the competitor does not carry an N-glycosylation site. Another set of assays takes advantage of the fact that at low temperatures, ATP-independent reversible peptide binding occurs, perhaps because of changes in the fluidity of the lipid environment.[60,61] Readout is through microsome associated radioactivity or through extent of photo-crosslinked peptide. An interesting feature for further biochemical and mechanistic studies is that these assays determine peptide affinities of TAP (directly or via competition experiments) as K_D-values in a bimolecular system.

Examination of the length selectivity of TAP has been hampered by two facts: (i) length preference and sequence preference are often hard to distinguish and (ii) demonstration of the structural integrity of peptides after translocation is a tedious but necessary task and thus results are often controversial. The most conclusive findings come from binding experiments with randomized peptides of different lengths[60] or direct translocation experiments and subsequent Edman-degradation of translocated peptides with an N-terminal N-glycosylation site and a C-terminal radiolabeled tyrosine-residue, to ensure that no potential degradation products are detected.[83,84] Human TAP prefers peptides with 8-16 residues,[60] although also considerably longer peptides could sometimes be translocated.[78] In the rat system, the situation was complicated by the fact that two allelic variants of different preferences exist, with rat TAP2[a] transporting an even broader peptide length range and with rat TAP2[u] more closely resembling the specificity of the human variant.[85] Hence, it can be concluded that the minimal length for translocation matches that optimal for MHC class I-binding (8-12 residues) but longer peptides can also be translocated, possibly followed by further trimming in the ER lumen.

The most profound effect on peptide specificity comes from the C-terminal residue of the translocated peptide. Again, a marked difference of the two allelic variants could be observed: rat TAP1.TAP2[u] and mouse TAP clearly select for aromatic and bulky hydrophobic residues, while rat TAP1.TAP2[a] and human TAP accept most residues but prefer basic and hydrophobic residues at the C terminus.[86] This pattern for human TAP was revealed by direct translocation assays, as well as by binding studies. Differences ranging over two orders of magnitude were observed in the binding assays, while differences in translocation based assays only reached up to five-fold. Also, when the same set of peptides was tested in both assay systems, remarkably similar results with respect to the relative order of preference were obtained, but again affinities from binding assays covered a much broader range. It could thus be speculated that the high concentrations used in the direct translocation assays, which are significantly above the expected K_M-values for high affinity peptides, lead to underestimation of the differences in affinity. Interestingly, preference for C-terminal residues matches that of MHC class I molecules, a fact that has been interpreted as a sign for coevolution of the molecules.[87]

In summary, the C-terminal residue of the substrate has the most important influence on peptide selection, followed by sequence positions at the N-terminal region of a nonamer peptide, while other positions seem to be only of minor importance. In particular, proline residues appeared to be strongly disfavored at position two by human TAP and position three by murine TAP, respectively.[88,89] Taking into account that free N- and C-termini are a strict prerequisite for affinity to TAP[90] this could be read as a contribution of the peptide backbone to peptide binding to TAP.

IV. PROTEINS ASSOCIATED WITH TAP

Are there any proteins or cofactors of the antigen presentation pathway that affect TAP function in vivo? Coimmunoprecipitation revealed the association of TAP heterodimers with peptide-free class I complexes[91] and indicated that this association is taking place via TAP1.[92] Further studies implied that calnexin, an ER lumenal chaperone, mediates heavy chain/β_2m dimerization and also subsequent TAP and class I complex association.[91] This association might facilitate the binding of antigenic peptides to the MHC molecules via local accumulation of peptides. Finally, peptide binding induces the release of MHC complexes from TAP and enables their transport to the Golgi compartment and the cell surface.[92] Recent evidence suggested that the TAP/class I interaction might even be necessary for peptide loading to MHC molecules, since a cell line expressing a point mutated heavy chain, which fails to bind to TAP, is impaired in peptide loading and antigen presentation.[93] By phenomenological characterization of the mutant cell line .220, in which also class I-TAP association and peptide loading are defective, the existence of yet unidentified MHC-linked gene products, which enhance or mediate TAP-MHC class I interaction, have been postulated.[94] Nevertheless, both TAP-mediated peptide transport in insect cells lacking MHC class I complexes and peptide binding by MHC molecules in TAP deficient cell lines can take place without association of these molecules,[36,95] thus indicating that the possibly enhancing effect might not be critical. Additionally, the existence of another protein has been reported, which is also specifically associated with TAP.[91,93] It was suggested that the interaction of heavy chain/β_2m dimers with TAP occurs via this novel, yet uncharacterized 48 kDa glycoprotein, tapasin, which can bind independently to TAP and class I-β_2m-calreticulin complexes. Interestingly, tapasin is absent from the mutant cell line .220.[96] Further studies will have to show the functional role of tapasin and other cofactors in antigen presentation.

V. TAP FUNCTION IN HUMAN PATHOGENESIS

The rat cim effect causes *tap* allele-dependent changes of the set of peptides presented on MHC class I molecules.[40,76] The genomic polymorphism and linkage disequilibrium in the human MHC-encoded antigen processing genes was analyzed with respect to several autoimmune diseases. Many studies speculated that polymorphic *tap* genes may or may not modulate the autoimmune response in insulin-dependent diabetes mellitus, multiple-sclerosis or rheumatoid arthritis.[97-101] However, although several human *tap1* and *tap2* alleles were identified,[38,39] no functional polymorphism for human TAP could be observed so far, using peptide transport assays in semi-permeabilized cells.[102] An inherited deficiency in the human TAP transporter was identified in two siblings suffering from recurrent respiratory bacterial infections. In these patients, the surface expression of MHC class I molecules was very low and the population of cytotoxic T lymphocytes, as well as the cytotoxicity of natural killer cells, was affected.[103]

Malignantly transformed or virus-infected cells are recognized by cytotoxic T lymphocytes due to a different set of peptide epitopes presented in complex with MHC class I molecules on the cell surface. Due to efficient pathways of antigen processing and presentation, tumors have evolved different strategies to escape immune surveillance, including a down-regulation of the *tap* genes.[104-108] Most interestingly, a functionally defective *tap1* allele (R659Q) was identified in human lung cancer by single strand conformational polymorphism.[62] The defective *tap1* allele causes loss of MHC class I surface expression and, thus, could result in tumor development.

Herpes simplex virus DNA codes for an immediate early protein, IE12 (ICP47), which was found to block antigen processing.[109] By specifically binding to TAP, this viral protein blocks peptide translocation into the ER lumen. Thereby, correct assembly and trafficking of MHC class I molecules is impaired.[110,111] It was demonstrated that recombinant, purified ICP47 inhibits peptide import into microsomes by human, but not by murine TAP.[112,113] The viral protein blocks TAP function by preventing peptide binding to TAP whereas ATP-binding appears unaffected by ICP47. Species-specificity of herpes simplex virus is achieved on the molecular level through a 100-fold higher affinity of ICP47 for human than for murine TAP. Since ICP47 represents the first natural inhibitor of an ABC transporter identified so far, it will be most interesting to investigate the structural and mechanistic requirements for TAP-ICP47 assembly. Synthetic TAP inhibitors designed on the basis of ICP47 would be highly useful in treatment of graft rejection or MHC class I related diseases. Vice versa, molecules disrupting the specific inhibitor complex of ICP47 and TAP would be a potent drug in the treatment of herpes virus infections.

REFERENCES

1. Bjorkman PJ, Strominger JL, Wiley DC. Crystallization and X-ray diffraction studies on the histocompatibility antigens HLA-A2 and HLA-A28 from human cell membranes. J Mol Biol 1985; 182: 205-210.
2. Bjorkman PJ, Saper MA, Samraoui B et al. Structure of the human class I histocompatibility antigen HLA-A2. Nature 1987; 329: 506-512.
3. Garrett TPJ, Saper MA, Bjorkman PJ et al. Specificity pockets for the side chains of peptide antigens in HLA-Aw68. Nature 1989; 342: 692-696.
4. Guo HC, Jardetzky TS, Garrett TPJ et al. Different length peptides bind to HLA-Aw68 similarly at their ends but bulge out in the middle. Nature 1992; 360: 364-366.
5. Collins EJ, Garboczi DN, Wiley DC. 3-dimensional structure of a peptide extending from one end of a class-I MHC binding-site. Nature 1994; 371: 626-629.
6. Glynne R, Powis SH, Beck S et al. A proteasome-related gene between the two ABC transporter loci in the class II region of the human MHC. Nature 1991; 353: 357-360.
7. Martinez CK, Monaco JJ. Homology of proteasome subunits to a major histocompatibility complex-linked *lmp* gene. Nature 1991; 353: 664-667.

8. Kelly A, Powis SH, Glynne R et al. Second proteasome-related gene in the human MHC class II region. Nature 1991; 353: 667-668.

9. Belich MP, Glynne RJ, Senger G et al. Proteasome components with reciprocal expression to that of the MHC-encoded Lmp proteins. Current Biol 1994; 4: 769-776.

10. Früh K, Gossen M, Wang KN et al. Displacement of housekeeping proteasome subunits by MHC-encoded Lmps—a newly discovered mechanism for modulating the multicatalytic proteinase complex. EMBO J 1994; 13: 3236-3244.

11. Akiyama KY, Yokota KY, Kagawa S et al. cDNA cloning and interferon down-regulation of proteasomal subunit X and subunit Y. Science 1994; 276: 1231-1234.

12. Driscoll J, Brown MG, Finley D et al. MHC-linked Lmp gene products specifically alter peptidase activities of the proteasome. Nature 1993; 365: 262-264.

13. Gaczynska M, Rock KL, Goldberg AL. Gamma-interferon and expression of MHC genes regulate peptide hydrolysis by proteasomes. Nature 1993; 365: 264-267.

14. Ehring B, Meyer TH, Eckerskorn C et al. Effects of major-histocompatibility-complex-encoded subunits on the peptidase and proteolytic activities of human 20S proteasomes—cleavage of proteins and antigenic peptides. Eur J Biochem 1996; 235: 404-415.

15. van Kaer L, Ashton-Rickardt PG, Eichelberger M et al. Altered peptidase and viral-specific T cell response in *lmp2* mutant mice. Immunity 1994; 1: 533-541.

16. Fehling HJ, Swat W, Laplace C et al. MHC class I expression in mice lacking the proteasome subunit Lmp7. Science 1994; 265: 1234-1237.

17. Rock KL, Gramm C, Rothstein L et al. Inhibitors of the proteasome block the degradation of most cell-proteins and the generation of peptides presented on MHC class-I molecules. Cell 1994; 78: 761-771.

18. Michalek MT, Grant EP, Gramm C et al. A role for the ubiquitin-dependent proteolytic pathway in MHC class I-restricted antigen presentation. Nature 1993; 363: 552-554.

19. Townsend A, Öhlen C, Bastin J et al. Association of class I major histocompatibility heavy and light chains induced by viral peptides. Nature 1989; 340: 443-448.

20. Cerundolo V, Alexander J, Anderson K et al. Presentation of viral antigen controlled by a gene in the major histocompatibility complex. Nature 1990; 345: 449-452.

21. Hosken NA, Bevan MJ. Defective presentation of endogenous antigen by a cell line expressing class I molecules. Science 1990; 248: 367-369.

22. Trowsdale J, Hanson I, Mockridge I et al. Sequences encoded in the class II region of the MHC related to the 'ABC' superfamily of transporters. Nature 1990; 348: 741-744.

23. Spies T, Bresnahan M, Bahram S et al. A gene in the human major histocompatibility complex class II region controlling the class I antigen presentation pathway. Nature 1990; 348: 744-747.

24. Deverson EV, Gow IR, Coadwell WJ et al. MHC class II region encoding proteins related to the multidrug resistance family of transmembrane transporters. Nature 1990; 348: 738-741.
25. Monaco JJ, Cho S, Attaya M. Transport protein genes in the murine MHC: possible implications for antigen processing. Science 1990; 250: 1723-1726.
26. Spies T, DeMars R. Restored expression of major histocompatibility class I molecules by gene transfer of a putative peptide transporter. Nature 1991; 351: 323-324.
27. Powis SJ, Townsend ARM, Deverson EV et al. Restoration of antigen presentation to the mutant cell line RMA-S by an MHC-linked transporter. Nature 1991; 354: 528-531.
28. Attaya M, Jameson S, Martinez CK et al. Ham-2 corrects the class I antigen-processing defect in RMA-S cells. Nature 1992; 355: 647-649.
29. Kelly AP, Powis SH, Kerr L-A et al. Assembly and function of the two ABC transporter proteins encoded in the human major histocompatibility complex. Nature 1992; 355: 641-644.
30. Kleijmeer M, Kelly A, Geuze HJ et al. Location of MHC-encoded transporters in the endoplasmic reticulum and cis-Golgi. Nature 1992; 357: 342-344.
31. Lévy F, Gabathuler R, Larsson R et al. ATP is required for in vitro assembly of MHC class I antigens but not for transfer of peptides across the ER membrane. Cell 1991; 67: 265-274.
32. Koppelman B, Zimmerman D, Walter P et al. Evidence for peptide transport across microsomal membranes. Proc Natl Acad Sci USA 1992; 89: 3908-3912.
33. Neefjes JJ, Momburg F, Hämmerling GJ. Selective and ATP-dependent translocation of peptides by the MHC-encoded transporter. Science 1993; 261: 769-771.
34. Androlewicz MJ, Anderson KS, Cresswell P. Evidence that transporter associated with antigen processing translocate a major histocompatibility complex cass I-binding peptide into the endoplasmic reticulum in an ATP-dependent manner. Proc Natl Acad Sci USA 1993; 90: 9130-9134.
35. Shepherd JC, Schumacher TN, P.G. A-R et al. *tap1*-dependent peptide translocation in vitro is ATP dependent and peptide selective. Cell 1993; 74: 577-84.
36. Meyer TH, van Endert PM, Uebel S et al. Functional expression and purification of the ABC transporter complex-associated with antigen-processing (TAP) in insect cells. FEBS Lett 1994; 351: 443-447.
37. Powis SH, Mockridge I, Kelly A et al. Polymorphism in a second ABC transporter gene located within the class II region of the human major histocompatibility complex. Proc Natl Acad Sci USA 1992; 89: 1463-1467.
38. Colonna M, Bresnahan M, Bahram S et al. Allelic variants of the human putative peptide transporter involved in antigen processing. Proc Natl Acad Sci USA 1992; 89: 3932-3936.

39. Carrington M, Colonna M, Spies T et al. Haplotypic variation of the transporter associated with antigen processing (TAP) genes and their extension of HLA class II region haplotypes. Immunogenetics 1993; 37: 266-273.

40. Powis SJ, Deverson EV, Coadwell WJ et al. Effect of polymorphism of an MHC-linked transporter on the peptides assembled in a class I molecule. Nature 1992; 357: 211-215.

41. Heemels MT, Schumacher TNM, Wonigeit K et al. Peptide translocation by variants of the transporter associated with antigen processing. Science 1993; 262: 2059-2063.

42. Wang P, Gyllner G, Kvist S. Selection and binding of peptides to human transporters associated with antigen processing and rat cim-a and -b. J Immunol 1996; 157: 213-220.

43. Higgins CF. ABC transporters: from microorganisms to man. Ann Rev Cell Biol 1992; 8: 67-113.

44. Walker JE, Saraste M, Runswick MJ et al. Distantly related sequences in the α- and β-subunits of ATP synthase, myosin, kinases and other ATP-requiring enzymes and a common nucleotide binding fold. EMBO J 1982; 1: 945-951.

45. Dassa E, Hofnung M. Sequence of gene *malG* in *E. coli* K12: homologies between integral membrane components from binding protein-dependent transport systems. EMBO J 1985; 4: 2287-2293.

46. Dassa E. Cellular localization of the MalG protein from the maltose transport system in *Escherichia coli* K12. Mol Gen Genetics 1990; 222: 33-36.

47. Cotten JF, Ostedgaard LS, Carson MR et al. Effect of cystic fibrosis-associated mutations in the 4th intracellular loop of cystic-fibrosis transmembrane conductance regulator. J Biol Chem 1996; 271: 21279-21284.

48. Hughes AL. Evolution of the ATP-binding-cassette transmembrane transporters of vertebrates. Mol Biol & Evol 1994; 11: 899-910.

49. Michaelis S, Berkower C. Sequence comparison of yeast ATP-binding cassette proteins. In: Cold Spring Harbor Symposia on Quantitative Biology, Volume LX, Cold Spring Harbor Laboratory Press, 1995.

50. Dean M, Allikmets R. Evolution of ATP-binding cassette transporter genes. Curr Opin Gen Dev 1995; 5: 779-785.

51. Chen CJ, Clark D, Ueda K et al. Genomic organization of the human multidrug resistance (*MDR1*) gene and origin of P-glycoproteins. J Biol Chem 1990; 265: 506-514.

52. Bult CJ, White O, Zhou L et al. Complete genome sequence of the methanogenic archaeon *Methanococcus jannaschii*. Science 1996; 273: 1058-1072.

53. Raymond M, Gros P, Whiteway M et al. Functional complementation of yeast Ste6 by a mammalian multidrug resistance *mdr* gene. Science 1992; 256: 232-234.

54. Volkman SK, Cowman AF, Wirth DF. Functional complementation of the ste6 gene of *Saccharomyces cerevisiae* with the *pfmdr1* gene of *Plasmodium falciparum*. Proc Natl Acad Sci USA 1995; 92: 8921-8925.

55. Ruetz S, Brault M, Kast C et al. Functional expression of the multidrug resistance-associated protein mrp in the yeast *Saccharomyces cerevisiae*. J Biol Chem 1996; 271: 4154-4160.

56. Müller KM, Ebensperger C, Tampé R. Nucleotide binding to the hydrophilic C-terminal domain of the transporter associated with antigen processing (TAP). J Biol Chem 1994; 269: 14032-14037.

57. Wang K, Früh K, Peterson PA et al. Nucleotide binding of the C-terminal domains of the major histocompatibility complex-encoded transporter expressed in *Drosophila melanogaster* cells. FEBS Lett 1994; 350: 337-341.

58. Russ G, Esquivel F, Yewdell JW et al. Assembly, intracellular localization, and nucleotide-binding properties of the human peptide transporters *tap1* and *tap2* expressed by recombinant vaccinia viruses. J Biol Chem 1995; 270: 21312-21318.

59. Meyer TH. ed. Functional expression and characterization of the MHC-encoded peptide transporter (TAP). Technical University Munich. 1996: Ph.D. Thesis.

60. van Endert PM, Tampé R, Meyer TH et al. A sequential model for peptide binding and transport by the transporters associated with antigen processing. Immunity 1994; 1: 491-500.

61. Uebel S, Meyer TH, Kraas W et al. Requirements for peptide binding to the human transporter associated with antigen-processing revealed by peptide scans and complex peptide libraries. J Biol Chem 1995; 270: 18512-18516.

62. Chen HL, Gabrilovic D, Tampé R et al. A functionally defective allele of *tap1* results in loss of MHC class I antigen presentation in a human lung cancer. Nature Gen 1996; 13: 210-213.

63. Holland I, Blight M. Structure and function of HlyB, the ABC transporter essential for haemolysin secretion from *Escherichia coli*. Biochim Biophis Acta (submitted for publication) 1996;

64. Dassa E, Muir S. Membrane topology of MalG, an inner membrane protein from the maltose transport system of *Escherichia coli*. Mol Microbiol 1993; 7: 29-38.

65. Ehrle R, Pick C, Ulrich R et al. Characterization of transmembrane domains 6, 7, and 8 of MalF by mutational analysis. J Bacteriol 1996; 178: 2255-2262.

66. Chang XB, Hou YX, Jensen TJ et al. Mapping of cystic-fibrosis transmembrane conductance regulator membrane topology by glycosylation site insertion. J Biol Chem 1994; 269: 18572-18575.

67. Geller D, Taglicht D, Edgar R et al. Comparative topology studies in *Saccharomyces cerevisiae* and in *Escherichia coli*—the N-terminal half of the yeast ABC protein Ste6. J Biol Chem 1996; 271: 13746-13753.

68. Kast C, Canfield V, Levenson R et al. Membrane topology of P-glycoprotein as determined by epitope insertion—transmembrane organization of the N-terminal domain of *MDR3*. Biochemistry 1995; 34: 4402-4411.

69. Kast C, Canfield V, Levenson R et al. Transmembrane organization of mouse P-glycoprotein determined by epitope insertion and immunofluorescence. J Biol Chem 1996; 271: 9240-9248.

70. Zhang J-T, Ling V. Study of membrane orientation and glycosylated extracellular loops of mouse P-glycoprotein by in vitro translation. J Biol Chem 1991; 266: 18224-18232.

71. Zhang J-T, Duthie M, Ling V. Membrane topology of the N-terminal half of the hamster P-glycoprotein molecule. J Biol Chem 1993; 268: 15101-15110.

72. Loo TW, Clarke DM. Membrane topology of a cysteine-less mutant of human P-glycoprotein. J Biol Chem 1995; 270: 843-848.

73. Loo TW, Clarke DM. Mutational analysis of the predicted first transmembrane segment of each homologous half of human P-glycoprotein suggests that they are symmetrically arranged in the membrane. J Biol Chem 1996; 271: 15414-15419.

74. van Iwaarden P, Pastore J, Konings W et al. Construction of a functional lactose permease devoid of cysteine residues. Biochemistry 1991; 30: 9595-9600.

75. Frillingos S, Kaback H. Cysteine-scanning mutagenesis of helix VI and the flanking hydrophilic domains in the lactose permease of *Escherichia coli*. Biochemistry 1996; 35: 5333-5338.

76. Powis SJ, Young LL, Joly E et al. The rat *cim* effect—tap allele-dependent changes in a class-I MHC anchor motif and evidence against C-terminal trimming of peptides in the ER. Immunity 1996; 4: 159-165.

77. Momburg F, Armandola EA, Post M et al. Residues in *tap2* peptide transporters controlling substrate-specificity. J Immunol 1996; 156: 1756-1763.

78. Androlewicz MJ, Cresswell P. Human transporters associated with antigen processing possess a promiscuous peptide binding site. Immunity 1994; 1: 7-14.

79. Nijenhuis M, Schmitt S, Armandola EA et al. Identification of a contact region for peptide on the *tap1* chain of the transporter associated with antigen processing. J Immunol 1996; 156: 2186-2195.

80. Armandola EA, Momburg F, Nijenhuis M et al. A point mutation in the human transporter associated with antigen-processing (*tap2*) alters the peptide-transport specificity. Eur J Immun 1996; 26: 1748-1755.

81. Greenberger LM. Major photoaffinity drug labeling sites for iodoaryl azidoprazosin in P-glycoprotein are within, or immediately C-terminal to, transmembrane domain 6 and domain 12. J Biol Chem 1993; 268: 11417-11425.

82. Morris DI, Greenberger LM, Bruggemann EP et al. Localization of the forskolin labeling sites to both halves of P-glycoprotein-similarity of the sites labeled by forskolin and prazosin. Mol Pharmacol 1994; 46: 329-337.

83. Koopmann JO, Post M, Neefjes JJ et al. Translocation of long peptides by transporters associated with antigen-processing (TAP). Eur J Immun 1996; 26: 1720-1728.

84. Momburg F, Roelse J, Hämmerling GJ et al. Peptide size selection by the major histocompatibility complex-encoded peptide transporter. J Exp Med 1994; 179: 1613-1623.

85. Heemels MT, Ploegh HL. Substrate-specificity of allelic variants of the TAP peptide transporter. Immunity 1994; 1: 775-784.

86. Momburg F, Roelse J, Howard JC et al. Selectivity of MHC-encoded peptide transporters from human, mouse and rat. Nature 1994; 367: 648-651.

87. Androlewicz MJ, Cresswell P. How selective is the transporter associated with antigen-processing. Immunity 1996; 5: 1-5.

88. Neisig A, Roelse J, Sijts AJA et al. Major differences in transporter associated with antigen presentation (TAP)-dependent translocation of MHC class I-presentable peptides and the effect of flanking sequences. J Immunol 1995; 154: 1273-1279.

89. Schumacher TN, Kantesaria DV, Heemels MT et al. Peptide length and sequence specificity of the mouse *tap1/tap2* translocator. J Exp Med 1994; 179: 533-540.

90. Androlewicz MJ, Ortmann B, van Endert PM et al. Characteristics of peptide and major histocompatibility complex class-I β(2)-microglobulin binding to the transporters associated with antigen-processing (*tap1* and *tap2*). Proc Natl Acad Sci USA 1994; 91: 12716-12720.

91. Ortmann B, Androlewicz MJ, Cresswell P. MHC class I/β 2-microglobulin complexes associate with TAP transporters before peptide binding. Nature 1994; 368: 864-867.

92. Suh WK, Cohen-Doyle MF, Früh K et al. Interaction of MHC class I molecules with the transporter associated with antigen processing. Science 1994; 264: 1322-1326.

93. Peace-Brewer AL, Tussey LG, Matsui M et al. A point mutation in HLA-A*0201 results in failure to bind the TAP complex and to present virus-derived peptides to CTL. Immunity 1996; 4: 505-514.

94. Grandea AG, Androlewicz MJ, Athwal RS et al. Dependence of peptide binding by MHC class-I molecules on their interaction with TAP. Science 1995; 270: 105-108.

95. Anderson K, Cresswell P, Gammon M et al. Endogenously synthesized peptide with an endoplasmic reticulum signal sequence sensitizes antigen processing mutant cells to class I-restricted cell-mediated lysis. J Exp Med 1991; 174: 489-492.

96. Sadasivan B, Lehner PJ, Ortmann B et al. Roles for calreticulin and a novel glycoprotein, tapasin, in the interaction of MHC class-I molecules with TAP. Immunity 1996; 5: 103-114.

97. Caillat ZS, Bertin E, Timsit J et al. Protection from insulin-dependent diabetes mellitus is linked to a peptide transporter gene. Eur J Immun 1993; 23: 1784-1788.

98. Wordsworth BP, Pile KD, Gibson K et al. Analysis of the MHC-encoded transporters *tap1* and *tap2* in rheumatoid arthritis: linkage with DR4 accounts for the association with a minor *tap2* allele. Tiss Antig 1993; 42: 153-5.

99. Moinsteisserenc H, Semana G, Alizadeh M et al. Tap2 gene polymorphism contributes to genetic susceptibility to multiple-sclerosis. Human Immunol 1995; 42: 195-202.

100. Vandevyver C, Stinissen P, Cassiman JJ et al. Tap-1 and Tap-2 transporter gene polymorphisms in multiple-sclerosis—no evidence for disease association with tap. J Neuroimmunol 1994; 54: 35-40.

101. Vandevyver C, Geusens P, Cassiman JJ et al. Peptide transporter genes (tap) polymorphisms and genetic susceptibility to rheumatoid-arthritis. British J Rheumatol 1995; 34: 207-214.

102. Obst R, Armandola EA, Nijenhuis M et al. TAP polymorphism does not influence transport of peptide variants in mice and humans. Eur J Immun 1995; 25: 2170-2176.

103. de la Salle H, Hanau D, Fricker D et al. Homozygous human TAP peptide transporter mutation in HLA class-I deficiency. Science 1994; 265: 237-241.

104. Cromme FV, Airey J, Heemels MT et al. Loss of transporter protein, encoded by the *tap1* gene, is highly correlated with loss of HLA expression in cervical carcinomas. J Exp Med 1994; 177: 505-509.

105. Kaklamanis L, Townsend A, Doussisanagnostopoulou IA et al. Loss of major histocompatibility complex-encoded transporter associated with antigen presentation (tap) in colorectal-cancer. American J Pathol 1994; 145: 505-509.

106. Rotemyehudar R, Winograd S, Sela S et al. Down-regulation of peptide transporter genes in cell-lines transformed with the highly oncogenic adenovirus-12. J Exp Med 1994; 180: 477-488.

107. Seliger B, Höhne A, Knuth A et al. Analysis of the major histocompatibility complex class I antigen presentation machinery in normal and malignant renal cells: evidence for deficiencies associated with transformation and progression. Cancer Res 1996; 56: 1756-1760.

108. Seliger B, Höhne A, Knuth A et al. Reduced membrane major histocompatibility complex class I density and stability in a subset of human renal cell carcinomas with low TAP and Lmp expression. Clin Cancer Res 1996; 2: 1427-1433.

109. York IA, Roop C, Andrews DW et al. A cytosolic herpes simplex virus protein inhibits antigen presentation to CD8+ T lymphocytes. Cell 1994; 77: 525-535.

110. Früh K, Ahn K, Djaballah H et al. A viral inhibitor of peptide transporters for antigen presentation. Nature 1995; 375: 415-418.

111. Hill A, Jugovic P, York I et al. Herpes simplex virus turns off the TAP to evade host immunity. Nature 1995; 375: 411-415.

112. Ahn K, Meyer TH, Uebel S et al. Molecular mechanism and species-specificity of TAP inhibition by herpes-simplex virus protein ICP47. EMBO J 1996; 15: 3247-3255.

113. Tomazin R, Hill AB, Jugovic P et al. Stable binding of the Herpes Simplex virus ICP47 protein to the peptide binding-site of TAP. EMBO J 1996; 15: 3256-3266.

114. Beck S, Kelly A, Radley E et al. DNA sequence analysis of 66 kb of the human MHC class II region encoding a cluster of genes for antigen processing. J Mol Biol 1992; 228: 433-441.

115. Chen CJ, Chin JE, Ueda K et al. Internal duplication and homology with bacterial transport proteins in the *MDR1* (P-glycoprotein) gene from multidrug-resistant human cells. Cell 1986; 47: 381-389.

116. van der Bliek AM, Kooiman PM, Schneider C et al. Sequence of *mdr3* cDNA encoding a human P-glycoprotein. Gene 1988; 71: 401-411.

117. Riordan JR, Rommens JM, Kerem B et al. Identification of the cystic fibrosis gene: cloning and characterization of complementary DNA. Science 1989; 245: 1066-73.

118. Mosser J, Douar AM, Sarde CO et al. Putative X-linked adrenoleukodystrophy gene shares unexpected homology with ABC transporters. Nature 1993; 361: 726-730.

119. Kamijo K, Kamijo T, Ueno I et al. Nucleotide sequence of the human 70 kDa peroxisomal membrane protein: a member of ATP-binding cassette transporters. Biochim Biophys Acta 1992; 1129: 323-327.

120. Thomas PM, Cote GJ, Wohllk N et al. Mutations in the sulfonylurea receptor gene in familial persistent hyperinsulinemic hypoglycemia of infancy. Science 1995; 268: 426-429.

121. McGrath JP, Varshavsky A. The yeast *STE6* gene encodes a homologue of the mammalian multidrug resistance P-glycoprotein. Nature 1989; 340: 400-404.

122. Kuchler K, Sterne RE, Thorner J. *Saccharomyces cerevisiae STE6* gene product: a novel pathway for protein export in eukaryotic cells. EMBO J 1989; 8: 3973-3984.

123. Dean M, Allikmets R, Gerrard B et al. Mapping and sequencing of two yeast genes belonging to the ATP-binding cassette superfamily. Yeast 1994; 10: 377-383.

124. Leighton J, Schatz G. An ABC transporter in the mitochondrial inner membrane is required for normal growth of yeast. EMBO J 1995; 14: 188-195.

125. Ortiz DF, Kreppel L, Speiser DM et al. Heavy metal tolerance in the fission yeast requires an ATP-binding cassette-type vacuolar membrane transporter. EMBO J 1992; 11: 3491-3491.

126. Hess J, Wels W, Vogel M et al. Nucleotide sequence of a plasmid-encoded haemolysin determinant and its comparison with a corresponding chromosomal haemolysin sequence. FEMS Microbiol Lett 1986; 34: 1-11.

127. Devereaux J, Haeberli P, Smithies O. A comprehensive set of sequence analysis programs for the VAX. Nucleic Acids Res 1984; 12: 387.

128. Felsenstein J. PHYLIP: phylogeny inference package (version 3.56). Cladistics 1989; 5: 164.
129. Claros MG, von Heijne G. Prediction of transmembrane segments in integral membrane proteins, and putative topologies, using several algorithms. CABIOS 1994; 10: 685-686.
130. Gottesmann MM, Hrycyna CA, Schoenlein PV et al. Genetic analysis of the multidrug transporter. Ann Rev Genet 1995; 29: 607-649.
131. Cheng SH, Gregory RJ, Marshall J et al. Defective intracellular transport and processing of CFTR is the molecular basis of most cystic fibrosis. Cell 1990; 63: 827-834.

CHAPTER 5

MAMMALIAN ABC TRANSPORTERS AND LEADERLESS SECRETION: FACTS AND SPECULATIONS

Yannick Hamon, Marie Françoise Luciani and Giovanna Chimini

I. INTRODUCTION

ABC transporters are one of the largest family of membrane transporters extremely conserved across evolution, from bacteria to mammals. Most of the members of the family function as ATP-dependent active transporters and the hallmark of the family lies in the presence of the ABC domain (ATP binding cassette).[1,2] A typical functional transporter consists of a symmetrical, pore-like structure across the membrane through which the substrate is thought to translocate. Although no universal model has been established, the binding and hydrolysis of ATP are thought to provide the energy for transport and the transmembrane (TMD) domains to determine the specificity for substrates. Unfortunately, the pleiotropism of the substrates transported by these proteins hampers an easy prediction of specificity of novel members. In fact, although their structural features are extremely well characterized, it has not been possible so far to correlate the presence of sequence motifs in the transporter to a particular substrate or substrate class. Substrate assignment has proven to be the most difficult task in the analysis of mammalian ABC transporters, even in cases when functional clues are provided by the existence of a pathological loss of function phenotype. In bacteria, on the contrary, ABC transporters were easily shown to handle a large variety of substrates.[3] In fact, bacterial ABC transporters move compounds in and out of the cell; they import sugars,[4-6] metal ions,[7,8] amino acids[9] and vitamins[10] and export pathogenic strain toxins such as hemolysin[11] to the extracellular medium. The ability of some bacterial and yeast members[12-14] to move proteins across membranes and in particular to promote

Unusual Secretory Pathways: From Bacteria to Man, edited by Karl Kuchler, Anna Rubartelli and Barry Holland. © 1997 R.G. Landes Company.

Table 5.1. Genomic distribution of the mammalian ABC transporters identified so far

	Man	Mouse
ABC-C	16p13	n.d.
ABC1	9q22-31	4A5-B3
ABC2	9q34	2A2-B
ABC7	Xq12-13	XC-D
ABC8	21q22.2-3	17A2-B
ALDp	Xq28	XA
ALDR	12q11-12	15E-F
CFTR	7q31	6A
MRP	16p13.1	n.d.
PGY1-3	7q21	5A2-A3
PMP70	1p22-21	3G-H1
TAP1-TAP2	6p21.3	17C
SUR	11p15.1-14	n.d.

Most of them have been cloned or mapped in mouse and in man (see text for individual refs.). The 21 EST showing similarity to ABC transporters are not listed (ref. 31).

secretion via nonclassical pathways prompted the hypothesis that they might as well be part of the alternative secretory pathway in mammalian cells. As described in other chapters of this book, this unusual way out from the cell concerns peptidic factors of high biological interest such as interleukin-1 and fibroblast growth factors.[15-17]

In this chapter we will summarize the general features of mammalian ABC transporters, in particular trying to highlight the problems and characteristics peculiar to mammalian systems. We will then focus on a novel ABC transporter, cloned and characterized in our lab,[18] which appears to be a good candidate for direct involvement in leaderless secretion.

II. ABC TRANSPORTERS IN MAMMALS

Until recently the number of ABC transporters identified in mammals had been surprisingly limited (Table 5.1). If we review the history of mammalian ABC transporters, it is easy to remark that in virtually all cases their identification proceeded from studies aimed to unravel the molecular basis of medically relevant phenotypes. The phenomenon of multidrug resistance developed by tumor cells is the paradigmatic example. The same holds true for the identification of the genes responsible for a number of genetic diseases, namely cystic fibrosis,[19] adrenoleukodystrophy[20] or hypoglycemia of the infancy,[21,22] all provoked by mutation or dysfunction of an ABC transporter. More recently and again by a positional cloning approach, the transporters associated with anti-gen presentation (TAP)[23] were identified, eventually providing the molecular answer to a long standing immunological dilemma.[24-27]

In the last years a number of groups, including ours, developed strategies targeted to the specific detection of novel members of the family.[18,28-30] These approaches, which will be discussed in detail later, together with the systematic analysis of the steadily increasing number of novel stretches of expressed sequences (EST), provided by the genome sequencing project, have allowed to highlight also in mammals a diversity of members of the ABC transporter family equaling that of bacteria.[31]

The quite recent availability of the complete sequence of the yeast genome has also provided another interesting analytical tool. It turns out indeed that the yeast genome possesses 28 ABC transporters which can be grouped in at least 5 structural subfamilies (Goffeau, personal communication).[32] All of them, with the exception of the PDR5 group are also represented in mammals. On the reverse direction, at least one of the to date identified mammalian structural groups, headed by ABC1, is not present in yeast. This is the first but might not be the sole example witnessing a functional diversification in this family of protein emerged with evolution to multi-cellularity. Indeed, the ABC1 structural family is lacking in yeast but is present in *C. elegans* and it would be certainly interesting to precisely assess its appearance in evolution.

III. THE STRUCTURE OF MAMMALIAN ABC TRANSPORTERS: RULES AND EXCEPTIONS

The functional ABC transporter shows a symmetrical four domain structure, composed by a membrane anchoring domain and an ABC fold repeated in tandem. The membrane anchoring domains, which in the vast majority of ABC transporters precede the ABC fold, are composed by six, as a rule, trans-membrane spanners delimiting the hydrophilic channel.[1,2,33] The ABC folds, the most conserved part of the molecule, lie on the cytosolic side of the membrane. Extremely conserved and diagnostic sequence motifs are present in the ABC folds; they are the two Walker motifs,[34] A and B, interacting directly with the nucleotide and its associated Mg^{2+} and the active transport signature (ATS), a short stretch of amino acids upstream to motif B whose function is so far undetermined (Fig. 5.1).[35,36]

This symmetry is, however, needed for function and results either from the synthesis of a single four domain polypeptide, internally repeated, and encoded by a single gene, or from the posttranslational association of two basic units or hemi-transporters (Fig. 5.2).[11] Although in bacteria examples exist of homodimerization, in mammals so far only heterodimerization of closely related hemi-transporters has been reported. This might reflect the nonequivalent

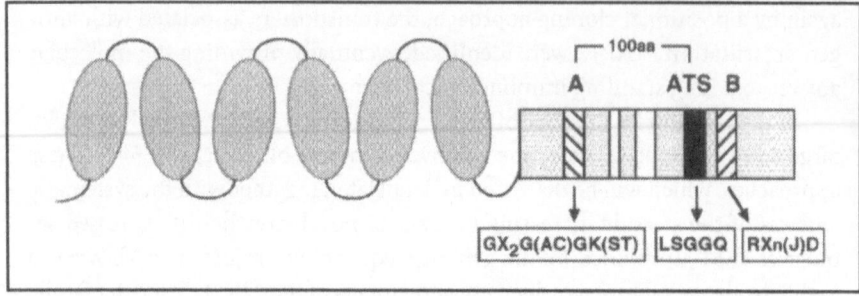

Fig. 5.1. Schematic diagram of the minimal structural subunit of an ABC transporter. The typical membrane anchoring domain consists of six membrane spanners followed by the ATP Binding Cassette. The latter is identified by the presence of a Walker motif A and B, whose distance is kept constant, and the Active Transport Signature (ATS). Internal diagnostic residues, conserved among the members of the family are shown by vertical bars. The one letter code for amino acids identification is used.

role of the first and second ABC folds in the final functional transporter, for which experimental evidence has been provided.[37-42] So far in mammals only six genes have been reported to encode hemitransporters. The two TAPs, whose physical interaction has been demonstrated, are present on the membrane of the endoplasmic reticulum, where they drive translocation of antigenic peptides.[43-47] The *PMP70*,[48] *ALDp*[20] and *ALDpR*[49] genes encode peroxisomal membrane proteins but their relative pattern of interaction is still an open question. Finally, we and others reported on the identification of ABC8, a mammalian homolog to *Drosophila white* gene, whose subcellular localization or dimerization partner is so far unknown. ABC8 is the first mammalian ABC transporter showing a reversed structure, that is the ABC fold preceding the TMD domain (J. Croop, unpublished results).[50,51]

Among the new expressed sequence tags belonging to ABC transporters some appear to be closely related to yeast *GCN20*,[52] reported to be involved in transcriptional regulation. They might, therefore, be examples of mammalian ABC transporters lacking the transmembrane domains. No clues are available so far on their molecular functioning, either as isolated units or in association with yet to be identified membrane anchoring structures.

Other variations on the archetypal structure have been reported for MRP[53] and the closely related cMOAT[54,55] and for SUR.[21] They all possess dissimilar membrane anchoring domains, the N-terminal containing more than the canonical six transmembrane spanners. In the case of SUR the proposed structural working model places the NH_2 terminus outside the cell with nine predicted TMD spanners before the first ABC fold. Four transmembrane helices separate the two ABC folds (Fig. 5.2).

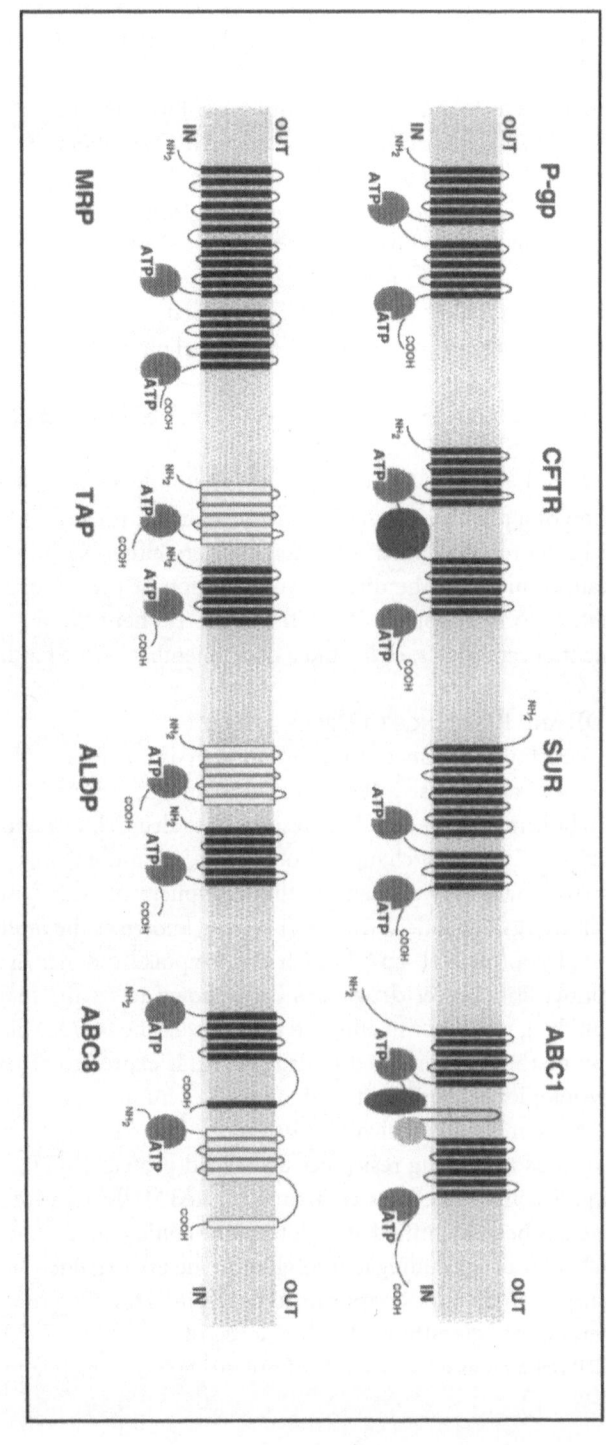

Fig. 5.2. Structural variations among mammalian ABC transporters. The dimerization partner for ALDp has not been assessed, although two other peroxisomal transporters have been identified (Pmp70 and ALDpR). In the case of ABC8, the mammalian homolog of *Drosophila white*, the partner of dimerization is unknown.

Although posttranscriptional regulation by phosphorylation has been reported for several ABC transporters,[56,57] the presence of additional structural domains exerting regulatory functions has been reported for CFTR and ABC1.[18,19] In *Dyctiostelium discoideum* an ABC transporter[58] with a unique structure has also been reported. This gene, tag B encodes a protein combining the features of both serine proteases and ABC transporters. The current model for its function implies processing and secretion of a putative peptidic substrate. These results suggest that several yet unpredictable structural variations are likely to exist and will soon be described along with the progress in the identification and characterization of novel mammalian genes.

IV. ABC TRANSPORTERS AND UNUSUAL SECRETORY PROCESSES

Although theoretically sound and appealing, no formal proof of the implication of a mammalian ABC transporter in the process of leaderless secretion has been provided. Most of the transporters, either because of their subcellular localization or for the direct demonstration of their substrate specificity, are unlikely to be involved. We will summarize here the data available for two potential candidates, *mdr1* and a novel member of the family, ABC1.

MDR AND RESISTANCE TO DRUGS

Resistance of cancer to chemotherapy is a complex phenomenon arising from several cell-based alterations that reduce accumulation of drugs, alter their metabolism, change cell cycle responses to drugs, block apoptosis in response to cytotoxic agents, change cytotoxic targets and enhance repair of drug induced damage. However, among this multiplicity of mechanisms, overexpression of an energy dependent transport system, known as the multidrug transporter, or P-glycoprotein (P-gp),[59] which blocks uptake and increases efflux of natural product anti-cancer drugs, has been shown to be the major mechanism of multidrug resistance in cultured cancer cell. The *mdr1*, which encodes in man genome the P-gp involved in drug efflux, is expressed at levels thought to be physiologically significant in about 50% of human cancers.

A second gene, which also belongs to the ABC transporter family, and is known as multidrug resistance associated protein (MRP),[60] was found to be amplified and overexpressed in drug selected MDR-negative cell lines. The MRP gene has been identified as a glutathione conjugate transporter[61] and it seems likely that drugs conjugated to glutathione are extruded from the cell by this transport system. MRP appears to be essential for drug resistance mediated by alterations in glutathione levels or levels of enzyme required for conjugation.[62] MRP behaves as a transporter of organic anions sensitive to probenecid.[63]

The first reports on cell lines acquiring simultaneous resistance to anti-cancer drugs appeared in the sixties.[64,65] Early studies identified both the physiological alteration and biochemical changes in these cells. These also pinpointed the pattern of multidrug resistance which included several different hydrophobic compounds. The identification of extrachromosomal elements, i.e., double minute chromosomes in highly resistant cells, suggested that the responsible gene was amplified in these cells and shortly afterwards the amplified segments were isolated and the gene encoding *mdr1* cloned and sequenced.[59] It was soon clear, however, that more than one *mdr* gene was present in man, mouse hamster and rat genome. Two genes are in fact present in man, *mdr1* and *mdr2*,[66] closely similar in sequence but certainly exerting different functions, since only *mdr1* is responsible for the drug resistance. In the mouse, three *mdr* genes are present, two of which encode functional multidrug transporters (*mdr1a* or 3[67,68] and *mdr1b* or 1[69]) and the third corresponds to human *mdr2*.[70] The function of MDR gene products as drug exporters has been extensively studied (see reference 71 for a general review). However, so far their physiological function as transporters has not been completely elucidated. Disruption of genes in mouse by the homologous recombination technique is still the more efficient way to assess their function in spite of a number of drawbacks, like embryonic lethality or gene redundancy. In the case of P-gps, knockouts have been generated for each of the three genes in addition to a double knockout in which both the genes *mdr1a* and *1b* have been disrupted.[72] The analysis of these animals has allowed to determine some or most of the speculated physiological functions for P-glycoproteins in mammals (Table 5.2).

The complete loss of *mdr1a* gene[73] has no apparent deleterious effect on mice, as long as they are not challenged with drugs normally transported by P-gp. Lethal doses and general toxicity are dramatically altered when the mice are challenged with drugs for which brain toxicity is limiting, such as vimentin. The body of results reported by Borst and coworkers illustrates the relevance of P-gp as an active extruder of amphipatic molecules that pass the blood brain barrier by diffusion. The analysis of double knockout mice is still limited but suggests that the drug transporting P-gps do not play essential roles in normal metabolism.[72] In particular no gross abnormality in corticosteroid metabolism, either during pregnancy or in bile formation, has been observed. Some of the suggested functions, like prenylcysteine or cholesterol transport, have not been analyzed in detail so far. As far as the secretion of leaderless proteins by *mdr1* is concerned, early reports by Young[74] on the ability to reconstitute interleukin-1β (IL-1β) secretion upon transfection of COS cells with *mdr1*, were never confirmed. In this respect, as it will be detailed later, the pharmacology of MDR does not parallel the known pharmacology of IL-1β secretion. On the other

Table 5.2. Suggested physiological functions for the P-glycoproteins involved in drug transport

Protection from exogenous toxins
Excretion of metabolites or toxins
Transport of steroid hormones
Extrusion of leaderless secretory proteins
Ion transport and cell volume regulation
Transport of prenylcysteine methyl esters
Intracellular vesicular transport of cholesterol

Modified from ref. 72.

hand both MRP and P-gp were shown to be able to complement *ste6* deficiency in yeast, implying that farnesylated peptides can be recognized and translocated as substrates by these transporters.[56,75]

mdr2 knockout mice[76] hinted at a novel function and supported a previously suggested transport model. Indeed *mdr2*[-/-] mice exhibit, as major defect, a progressive hepatic sclerosis, consequent to a reduced excretion rate of phosphatidylcholine in the bile. A role for *mdr2* in the transport or the flipping of phosphatidylcholine could be at the basis of the defect and recent reconstitution of this translocase activity in yeast vesicles supports this hypothesis.[77,78]

THE CASE OF ABC1

A targeted cloning approach

The strategy that we employed to isolate novel members of this family of ATPases was based on the observation that they all share regularly spaced sequence motifs located in the ATP binding cassette (Fig. 5.1). Motif A and ATS, whose spacing varies little among the already known members, were chosen as targets for amplification. Mixed oligonucleotide primed amplification was carried out on RNA from cells of the monocyte/macrophage lineage and resulted in a major amplification band of the expected 340 bp length. We expected this band to contain a mixture of sequences, resulting from the amplification of the diverse members of the family expressed in the original sample. The cloning of the amplification product led to the isolation of individual clones and their analysis to the identification of several bona fide ABC cassettes. These sequences, in fact, encoded an open reading frame in the correct orientation between the two primers featuring the conserved diagnostic internal residues and showed significant homology with already known ABC folds from members of the family.[35] Further systematic analysis of expression pattern, genomic mapping and full-length cloning was carried out and led to the characterization of 5 novel ABC transporters.[18,49,50]

One of them, ABC1, is described in detail since it shows a peculiarity of structure, and the search for its function has revealed several unexpected features. Although their significance has not been completely elucidated yet, it seems likely that ABC1 and/or closely related proteins might play a role in the process of secretion of leaderless proteins in mammalian cells.

A novel structural group

The ABC1 gene maps to human chromosome 9q22-31 and mouse chromosome 4,A5-B3, and encodes a large internally symmetrical protein (predicted molecular weight of 220 kDa) which shows all the structural features typical for ABC transporters. Two similar halves, each containing a set of six putative transmembrane spanners and a nucleotide binding domain (the ABC cassette), are repeated in tandem, thus providing the complete structure for a functional channel through the membrane. Its structure is, however, unique among ABC transporters since between the two halves an extremely long putative regulatory domain interrupted by a stretch of hydrophobic residues is present (Fig. 5.3). Whereas the presence of a regulatory domain has been already reported for CFTR,[19] so far none of the known transporters shows an extra hydrophobic segment potentially able to interact with the membrane or with an additional partner (the substrate or an intermediate substrate binding protein) in a regulated fashion. Apparently, however, this extra hydrophobic segment does not perturb the overall membrane topology of the transporter with respect to the generally accepted topological model. Both ABC cassettes are in fact oriented in the cytosolic side on the membrane as assessed in protease sensitivity assays on in vitro-translated, membrane-associated chimeric ABC1 proteins (Hamon et al, unpublished results). This structure is shared by ABC2 and the recently reported ABC-C (or ABC 3) gene product.[7,809] ABC1 was cloned from mouse macrophages cDNA (P388DI) and appears to be widely expressed in adult tissues and during embryonic development. Its expression is, however, not restricted to macrophage since its transcript is detectable in several established cell lines. A partial clone of human origin was also sequenced and the sequence conservation is higher than 95% at the DNA and protein level—a feature which is not uncommon among ABC transporters.

ABC1 and cell death: Lessons from evolution

Three lines of evidence support the implication of ABC1 in the genetic control of programmed cell death (PCD).[81]
- the morphological correlation, which established the presence of ABC1 transcript during development in areas of PCD, such as the interdigital web at E13,5 (Fig. 5.4).[81] There ABC1 is expressed in cells of macrophage origin, as defined by positivity with the pan-macrophage marker F4/80,[83]

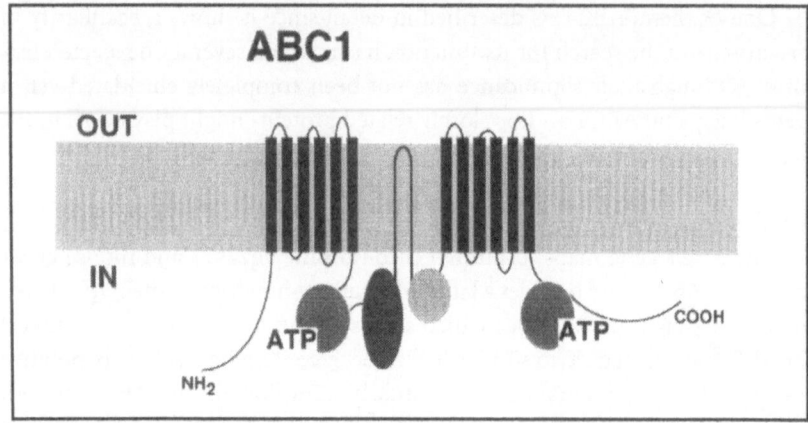

Fig. 5.3. Predicted membrane topology for ABC1 and related transporters. The interaction with the membrane of the hydrophobic stretch of amino acid splitting the regulatory domain is based on computer predictions.

engaged in the engulfment of apoptotic corpses. The same correlation is present in another model of apoptosis—the thymus induced to massive death by corticosteroid treatment.[84]

- the requirement for ABC1 function during engulfment, which was established by the dramatic impairment of the ability of peritoneal macrophages to phagocytose apoptotic thymocytes after antibody-mediated steric blockade of ABC1.

- finally the evolutionary conservation which established ABC1 as a potential mammalian homolog of *ced-7* gene fom *C. elegans*.[85] Indeed, sequence comparison highlighted approx. 30% identity at the aminoacid level between ABC1 and an ABC transporter on *C. elegans* chromosome III, further identified by Horvitz et al as *ced-7*.[86] It has to be noted that the structural similarities between ABC1 and *ced-7* include the conservation of the extra-hydrophobic segment. This is remarkable since, as already mentioned, none among the ABC transporters present in yeast shares this structural feature (André Goffeau personal communication).

Ced-7 belongs to the group of mutations identified in the nematode as responsible for the engulfment of corps generated by cell death.[85] As widely known, *C. elegans* is today one of the most useful model systems for the understanding of the cell death process.[87] During normal development of this animal, 131 cells out of the 1090 generated nuclei undergo programmed cell death. These events are precisely controlled and the pattern of their occurrence is known in detail.[87,88]

Morphologically, the death process can be split into several phases which correspond to precise genetic information (Fig. 5.5). The cell committed to die

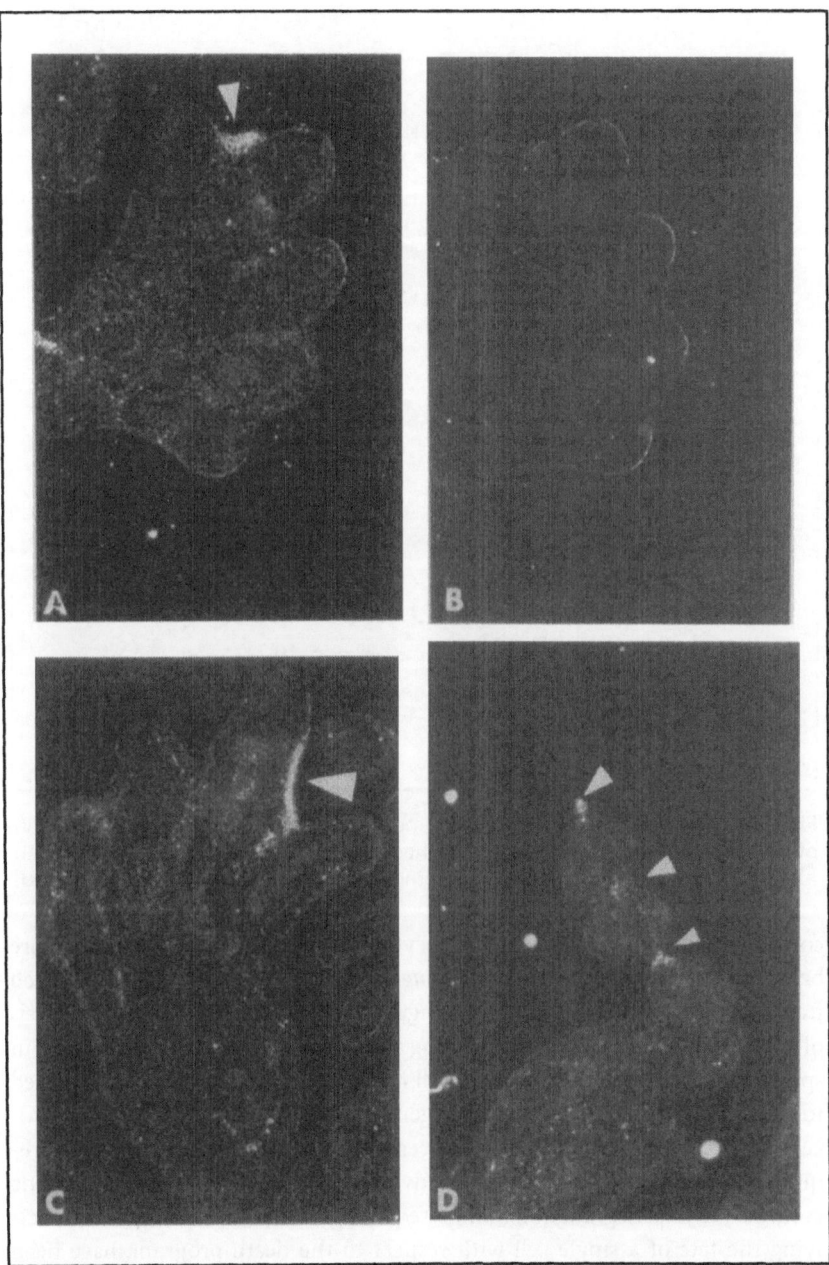

Fig. 5.4. Expression of ABC1 transcript in the interdigital web during limb development. Dark field images of RNA in situ hybridization analysis with the ABC1 specific probe on sagittal (A-C) and cross sections (D) from mouse hind limb bud at E13 (A) or E14 (C, D). Negative control hybridization is shown in B.

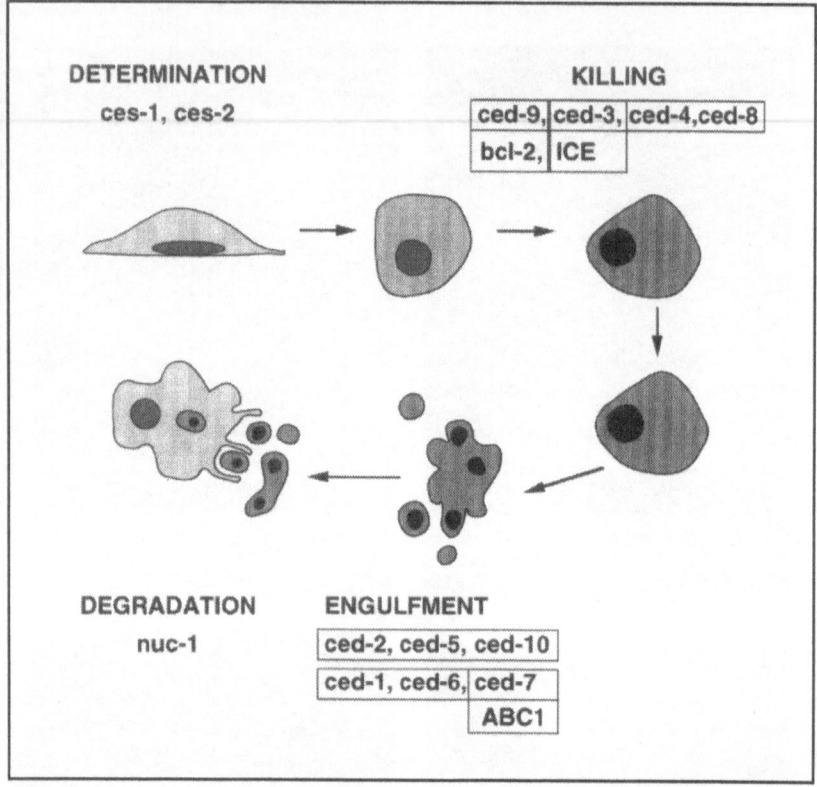

Fig. 5.5. The genetic control of programmed cell death. Schematic representation of the phases of the death process and associated genes. *ced* genes have been identified in *C. elegans*. Mammalian homologs, when known, are also indicated.

condenses, rounds up and undergoes cytoplasmic contraction (*killing*) before being engulfed by neighboring cells (*engulfment*). Once engulfed, the dead cell fragments and is degraded within the engulfing cell (*degradation*). By the analysis of chemically induced mutants showing abnormality in the cell death program specific sets of genes (called *ced* for *c*ell *d*eath abnormal) have been associated to each of these phases. Four of these genes control the onset of the death process (*ced-3, ced-4, ced-9* and more recently *ced-8*) whereas six others are required during the engulfment of the dying cells by its neighbors. A last gene controls the final degradation of ingested corpses. Two additional genes, specifying the fate of a single cell with respect to the death program, have been identified (*determination*).

Most interestingly, some of these genes show structural and functional similarities to genes that act in cell death in vertebrates. This is the case, for instance, for two genes regulating and executing the death sentence: *ced-9*, which acts to protect from death cells that should survive; and *ced-3* which is required for death to occur.[89-91]

Either gain of function or overexpression in the *ced-9* gene leads to survival of cells normally committed to death, whereas loss of function mutants causes extensive increase of death during development leading to embryonic lethality. Similar protective effects have been associated to vertebrate *bcl-2* gene, which shows a striking sequence similarity to the *C. elegans ced-9*. In addition, these two genes appear to be functionally interchangeable since *ced-9* mutants can be rescued by the vertebrate homolog. This finding strongly suggests that the genetic programs for cell death in nematodes and mammals share common molecular mechanism and is witness of an ancient evolutionary origin predating separation of nematodes and vertebrates. Further support of this evolutionary conservation is the identification of *the ced-3* gene as a structural and functional homolog of the mammalian interleukin-1β-converting enzyme (ICE) and related proteases.[89,92]

As far as engulfment phase is concerned, 6 genes (*ced-1, ced-2, ced-5, ced-6, ced-7 and ced-10*) whose loss of function mutants shows the presence of uningested cell corpses have been identified.[85] From the study of double mutants with additive phenotype they can be assigned to two subsets controlling different and partially redundant processes, both able to cause a cell corpse to be engulfed. One engulfment process involves the genes *ced-2,ced-5* and *ced-10* whereas the other involves *ced-1, ced-6* and *ced-7*. Even when one of these processes is impaired, the other can still cause the engulfment of cell corpses. If both processes are completely blocked, engulfment does not occur, although corpses might still be lost by lysis or detachment due to their intrinsic fragility. The *ced-2, ced-5, ced-10* pathway appears to control also more general aspects of cell spreading and migration (Hengartner and Horvitz, personal communication). So far only three of these genes have been cloned and sequenced, among them, *ced-7* (Horvitz and Wu, unpublished results). *ced-7* encodes an ABC transporter which shows all the classical structural features of this family of proteins and an additional large regulatory domain split by an extra-hydrophobic segment. These findings and the morphological and functional data strongly suggest ABC1 as a mouse homolog of *ced-7*, once again supporting the hypothesis of a common genetic control of death programs conserved across broad evolutionary distances.

ABC1 as an anion transporter

Aiming at acquiring information on the molecular function of ABC1, we analyzed its activity as a channel or transporter after expression in *Xenopus* oocytes. By analogy with other known ABC transporters, we analyzed whether novel currents or ionic fluxes were specifically generated by the expression of

ABC1. Our results suggest that ABC1 behaves in frog oocytes as an electroneutral anionic transporter as judged by iodide efflux measurements. ABC1 function as an anion exchanger is sensitive to several drugs (glibenclamide, bromosulfophtaleine and DIDS), resulting in an ABC1-specific pharmacological profile, and can be regulated by a cAMP-dependent pathway (Fig. 5.6).[93]

As a further confirmation of the functional conservation across evolution it is worth noting that *ced-7* expressed in *Xenopus* oocytes does generate a similar flux, with a similar pharmacological profile (Hamon et al, unpublished observation).

Interestingly enough, we could detect an anion flux with an identical pharmacological profile in thioglycollate-induced murine peritoneal macrophages, i.e., inflammatory cells expressing ABC1 (Luciani, unpublished results). Preliminary results show that the drug sensitivity of ABC1-dependent fluxes and the secretion of IL-1β are consistently superposable. No effect is seen on the secretion of control cytokines like TNF and IL-6. The working hypothesis, therefore, is that ABC1 or a closely related protein share the same pharmacology functions along the alternative secretory pathway for IL-1β.

INTERLEUKIN-1β AND THE LEADERLESS SECRETORY PATHWAY

Understanding interleukin 1β secretion is a long-standing challenge in immunology.[94] Interleukin 1 is the prototypic multifunctional cytokine able to affect nearly every cell type, often in concert with other cytokines or mediators. IL-1β is a highly inflammatory cytokine and the margin between the benefit and clinical toxicity in humans is exceedingly narrow. Therefore, the development of agents able to control the production, secretion and/or activity of IL-1β is likely to have an impact on clinical medicine. Indeed, the whole process of IL-1β production is a tightly regulated event. Along the pathway several steps are highly controlled: gene expression, synthesis and secretion; the regulation extends also to surface receptors, soluble receptors and receptor antagonists. Activated monocytes or macrophages are the primary source of this cytokine, which is synthesized after a wide variety of stimuli, the most potent being LPS or other bacterial cell wall products. Perturbation of ionic equilibria across cell membranes has also been reported as pivotal in the control of IL-1β secretion.[95] Indeed a panel of drugs, already known or newly developed blockers of anion transport, have been shown to inhibit the secretion of this cytokine by a mechanism which is not dependent on the lowering of intracellular pH.[96-98]

As far as the process of secretion is concerned, the first remarkable point is the unusual cellular route that it takes. In fact, IL-1β is first synthesized as a cytosolic inactive precursor (pro-IL-1) of 31 kDa lacking a signal peptide, which hampers its access to the ER and the classical exocytic pathway.[15] A large body of evidence has been provided to exclude its secretion via the classical exocytic route. After synthesis, maturation and secretion (implying the translocation of

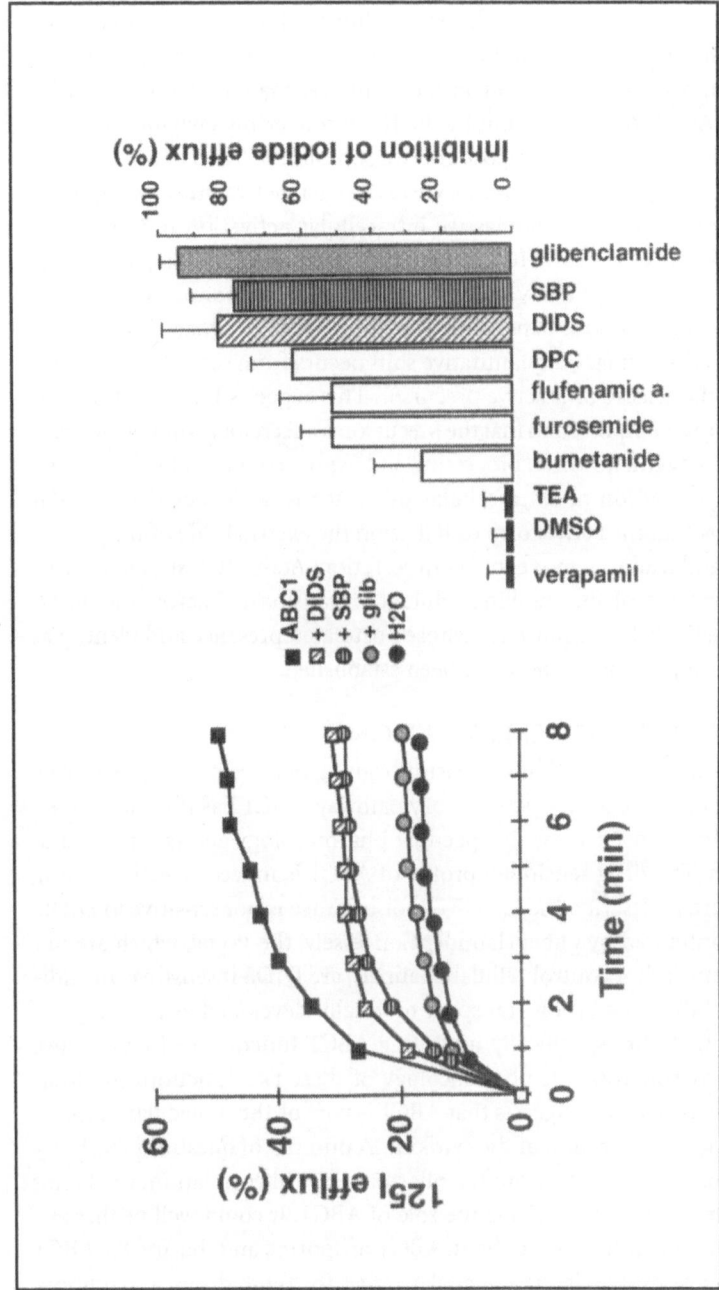

Fig. 5.6. ABC1 is an anionic transporter with a specific pharmacological profile. The expression of ABC1 in *Xenopus laevis* oocytes generates a specific anion flux as detected by iodide efflux assays (ref. 93). A. Kinetic profile of iodide efflux from oocytes injected with water or ABC1. The effect of treatment with drugs is shown. B. Histograms comparing the inhibitory effects of different compounds on ABC1 generated iodide efflux from oocytes. SBP, sulfobromophthalein; DIDS, 4,4'-diisothiocyanostilbene-2, 2'-disulfonic acid; DPC, diphenylamine-2-carboxylic acid; TEA, tetraethyl ammonium.

a cell membrane to gain access to the extra cellular milieu) have to occur. Unfortunately the pathway is still not precisely defined. The processing and secretion events appear so far to occur in a cotemporal manner. Cleavage of the precursor at Asp-116 is carried out by the IL-1β converting enzyme (ICE), an heterodimeric thiolprotease composed of two subunits, p20 and p10. ICE which is constitutively synthesized by monocytes as a p45 inactive precursor appears to be mainly cytosolic.[99-101] Its massive intracellular activation in transfected cells leads to cell death, and no ICE activity is detectable in physiological conditions of IL-1β secretion. Finally, the treatment of human monocytes with ICE inhibitors results in a dose-dependent reduction of IL-1β processing without affecting secretion. In fact, a quantitative shift occurs from secretion of the active form to the release of inactive precursor. This suggests that secretion and maturation are cotemporal and that the precursor is a secretory competent form. In the mouse system, however, processing and export could not be dissociated since the accumulation of extra-cellular precursor is far less evident. Recent results suggest that the active form of ICE is on the external side of the plasma membrane, and lead to a model of IL-1β secretion. Active ICE should process the precursor form of the cytokine, while it is translocated across the membrane by a dedicated transporter,[102] whose postulated presence and identity as an ABC transporter has not however been established.

V. CONCLUSION AND SPECULATION

Several features of ABC1 are at least intriguing and seem to suggest a role for this protein in the leaderless secretory pathway. ABC1 has the characteristics of an anion transporter with a peculiar pharmacology and is expressed in macrophages. The drug sensitivity profile of ABC1 is indeed selective among ABC transporters described so far. CFTR for instance is not sensitive to DIDS, while being inhibited by glibenclamide. Conversely, the P-gps, which are not chloride channels, but control cellular channels, are DIDS insensitive. In addition, several inhibitors of ion transport originally developed as affecting the secretion of IL-1β are specifically inhibiting ABC1 function in frog oocytes. The exquisitely superposable pharmacology of these two functions in physiological systems strongly suggests that ABC1 is part of the molecular machinery required for the secretion of the cytokine. A number of questions are, however, still to be solved. Apart from the pharmacological correlation, we do not have so far any evidence clarifying the role of ABC1. It could well be that another molecule, likely however to be an ABC transporter, and sharing the ABC1 pharmacological profile, is the molecular target for these drugs. In addition, anion transport could affect indirectly the secretory pathway and trigger only the function of other yet to be identified molecular entities.

REFERENCES

1. Ames GF-L. Bacterial periplasmic transport system: structure, mechanism and evolution. Ann Rev Biochem 1986; 55: 377-425.
2. Higgins CF. ABC transporters: from microorganisms to man. Ann Rev Cell Biol 1992; 8: 67-113.
3. Fath MJ, Kolter R. ABC transporters: Bacterial exporters. Microbiol Rev 1993; 57: 995-1017.
4. Gilson E, Nikaido H, Hofnung, M Sequence of the *malK* gene in *E. coli* K12. Nucleic Acids Res 1982; 10: 7449-7458.
5. Froshauer S, Beckwith J. The nucleotide sequence of the gene for *malF* protein, an inner membrane component of the maltose transport system of *Escherichia coli.* J Biol Chem 1984; 259: 10896-10903.
6. Dassa E, Hofnung M. Sequence of gene *MalG* in *E coli* K12: homologies between integral membrane components from binding protein-dependent transport systems. EMBO J 1985; 4: 287-293.
7. Coulton JW, Mason P, Alatt DD. *fhuC* and *fhuD* genes for iron (III)-ferrichrome transport into *Escherichia coli* K-12. J Bacteriol 1987; 169: 3844-3849.
8. Kostler W, Braun V. Iron hydroxamate transport of *Escherichia coli*: nucleotide sequence of the *fhuB* gene and identification of the protein. Mol Gen Genet 1986; 204: 435-442.
9. Higgins CF, Haag PD, Nikaido K et al. Complete nucleotide sequence and identification of membrane components of the histidine transport operon of *S. typhimurium.* Nature 1982; 298: 723-727.
10. Friedrich MJ, Deveaux LC, Kadner RJ. Nucleotide sequence of the *btu*CED genes involved in vitamin B_{12} transport in *Escherichia coli* an homology with components of periplasmic-binding-protein-dependent transport systems. J Bacteriol 1986; 167: 928-934.
11. Felmlee T, Pellett S, Welch RA. Nucleotide sequence of an *Escherichia coli* chromosomal haemolysin. J Bacteriol 1985; 163: 94-105.
12. Kuchler K. Unusual routes of secretion: the easy way out. Trends in Cell Biol 1993; 3: 421-426.
13. Kuchler K, Sterne RE, Thorner J. *Saccharomyces cerevisiae STE6* gene product: a novel pathway for protein export in eukaryotic cells. EMBO J 1989; 8: 3973-3984.
14. Kuchler K, Thorner J. Secretion of peptides and proteins lacking hydrophobic signal sequences: the role of adenosine triphosphate-driven membrane translocators. Endocrine Rev 1992; 13: 499-514.
15. Rubartelli A, Cozzolino F, Talio M et al. A novel secretory pathway for interleukin-1β, a protein lacking a signal sequence. EMBO J 1990; 9: 1503-1510.
16. Rubartelli A, Bajetto A, Allavena G et al. Post-translationnal regulation of interleukin-1β secretion. Cytokine 1993; 5: 117-124.
17. Muesch A, Hartmann E, Rohde K et al. A novel pathway for secretory proteins? Trends Cell Biol 1990; 15: 86-88.

18. Luciani MF, Denizot F, Savary S et al. Cloning of two novel ABC transporters mapping on human chromosome 9. Genomics 1994; 21: 150-159.
19. Riordan JR, Rommens JM, Kerem B-S et al. Identification of the Cystic Fibrosis gene: cloning and characterization of complementary DNA. Science 1989; 245: 1066-1073.
20. Mosser J, Douar A-M, Sarde CO et al. Putative X-linked adrenoleucodystrophy gene shares unexpected homology with ABC transporters. Nature 1993; 361: 726-730.
21. Aguilar-Bryan L, Nichols CG, Wechsler SW et al. Cloning of the b cell high-affinity sulfonylurea receptor: a regulator of insulin secretion. Science 1995; 268: 423-426.
22. Thomas PM, Cote GJ, Wohllk N et al. Mutations in the sulfonylurea receptor gene in familial persistent hyperinsulinic hypoglycemia of infancy. Science 1995; 268: 426-429.
23. Kelly A, Powis SH, Kerr L-A et al. Assembly and function of the two ABC transporter proteins encoded in the human major histocompatibility complex. Nature 1992; 355: 641-644.
24. Deverson EV, Gow IR, Coadwell WJ et al. MHC class II region encoding proteins related to the multidrug resistance family of transmembrane transporters. Nature 1990; 348: 738-741.
25. Monaco JJ, Cho S, Attaya M. Transport protein genes in the murine MHC: possible implications for antigen processing. Science 1990; 250: 1723-1726.
26. Spies T, Bresnahan M, Bahram S et al. A gene in the human major histocompatibility complex class II region controlling the class I antigen processing pathway. Nature 1990; 348: 744-747.
27. Trowsdale J, Hanson I, Mockridge I et al. Sequence encoded in the class II region of the MHC related to the 'ABC' superfamily of transporters. Nature 1990; 348: 741-744.
28. Leighton J, Schatz G. An ABC transporter in the mitochondrial inner membrane is required for normal growth of yeast. EMBO J 1995; 14: 188-195.
29. Kuchler K, Göransson HM, Viswanathan MN, Thorner J. Dedicated transporters for peptide export and intercompartmental traffic in the yeast *Saccharomyces cerevisiae*. In: Cold Spring Harbor Symposia on Quantitative Bology, Volume LVII, Cold Spring Harbor: Cold Spring Harbor Laboratory Press, 1992; p. 579-592.
30. Dean M, Allikmets R, Gerrard B et al. Mapping and sequencing of two yeast genes belonging to the ATP-binding cassette superfamily. Yeast 1994; 10: 377-383.
31. Allikmets R, Gerrard B, Dean M. Characterization and mapping of 21 new human genes from the ABC superfamily using the expressed sequence tags database. Hum Mol Genet 1996; 5:1669-53.
32. Michaelis S, Berkower C. Sequence comparison of yeast ATP-binding cassette proteins. In: *Cold Spring Harbor Symposia on Quantitative Biology, Volume LX*, Cold Spring Harbor Laboratory Press, 1995.

33. Kast C, Canfield V, Levenson R et al. Transmembrane organization of mouse P-glycoprotein determined by epitope insertion and immunofluorescence. J Biol Chem 1996; 271: 9240-9248.

34. Walker JE, Saraste M, Runswick MJ et al. Distantly related sequences in the a and b subunits of ATP synthase, myosin, kinases and other ATP-requiring enzymes and a common nucleotide binding fold. EMBO J 1982; 8: 945-951.

35. Mimura CS, Holdbrook SR, Ames GF-L. Structural model of the nucleotide-binding conserved component of periplasmic permeases. Proc Natl Acad Sci USA 1991; 88: 84-88.

36. Hyde SC, Emsley P, Hartshorn MJ et al. Structural model of ATP-binding proteins associated with cystic fibrosis, multidrug resistance and bacterial transport. Nature 1990; 346: 362-365.

37. Russ G, Esquivel F, Yewdell JW et al. Assembly, intracellular localization, and nucleotide binding properties of the human peptide transporters *TAP1* and *TAP2* expressed by recombinant vaccinia viruses. J Biol Chem 1995; 270: 21312-21318.

38. Berkower C, Michaelis S. Mutational analysis of the yeast a-factor transporter *STE6*, a member of the ATP binding cassette (ABC) protein superfamily. EMBO J 1991; 10: 3777-3785.

39. Müller M, Bakos E, Welker E et al. Altered drug-stimulated ATPase activity in mutants of the human multidrug resistance protein. J Biol Chem 1996; 271: 1877-1883.

40. Beaudet L, Gros P. Functional dissection of P-glycoprotein nucleotide-binding domains in chimeric and mutant proteins. Modulation of drug resistance profiles. J Biol Chem 1995; 270: 17159-17170.

41. Urbatsch IL, Sankaran B, Bhagat S et al. Both P-glycoprotein nucleotide-binding sites are catalytically active. J Biol Chem 1995; 270: 26956-26961.

42. Carson MR, Travis SM, Welsh, MJ. The two nucleotide-binding domains of cystic fibrosis transmembrane conductance regulator (CFTR) have distinct functions in controlling channel activity. J Biol Chem 1995; 270: 1711-1717.

43. Androlewicz MJ, Anderson KS, Cresswell, P. Evidence that transporters associated with antigen processing translocate a major histocompatibility complex class I-binding peptide into the endoplasmic reticulum in an ATP-dependent manner. Proc Natl Acad Sci USA 1993; 90: 9130-9134.

44. Androlewicz MJ, Cresswell P. Human transporters associated with antigen processing possess a promiscuous peptide-binding binding site. Immunity 1994; 1: 7-14,

45. Androlewicz MJ, Ortmann B, Van Endert PM et al. Characteristics of peptide and major histocompatibility complex class I/b_2-microglobulin binding to the transporters associated with antigen processing (*TAP1* and *TAP2*). Proc Natl Acad Sci USA 1994; 91: 12716-12720.

46. Androlewicz MJ, Cresswell P. How selective is the transporter associated with antigen processing? Immunity 1996; 5: 1-5.

47. Neefjes JJ, Momburg F, Hämmerling GJ. Selective and ATP-dependent translocation of peptides by the MHC-encoded transporter. Science 1993; 261: 769-771.
48. Kamijo K, Taketani S, Yokota S et al. the 70 kDa peroxisomal membrane protein is a member of the MDR (P-glycoprotein)-related ATP-binding protein superfamily. J Biol Chem 1993; 265: 4534-4540.
49. Lombard-Platet G, Savary S, Sarde CO et al. A close relative of the adrenoleukodystrophy (ALD) gene codes for a peroxisomal protein with a specific expression pattern. Proc Natl Acad Sci USA 1996; 93: 1265-1269.
50. Savary S, Denizot F, Luciani MF et al. Molecular cloning of a mammalian ABC transporter homologous to *Drosophila white* gene. Mammal Genome 1996; 7: 673-676.
51. Chen H, Rossier C, Lalioti MD et al. Cloning of the cDNA for a human homolog of the *Drosophila white* gene and mapping to chromosome 21q22.3. Am J Hum Genet 1996; 59: 66-75.
52. Vazquez de Aldana CR, Marton MJ, Hinnebusch AG. *GCN20.* a novel ATP binding cassette protein, and *GCN1* reside in a complex that mediates activation of the eIF-2a kinase *GCN2* in amino acid-starved cells. EMBO J 1995; 14: 3184-3199.
53. Loe DW, Deeley RG, Cole SPC. Biology of the multidrug resistance-associated protein, MRP. Eur J Cancer [A] 1996; 32A: 945-957.
54. Paulusma CC, Bosma PJ, Zaman GJR et al. Congenital jaundice in rats with a mutation in a multidrug resistance-associated protein gene. Science 1996; 271: 1126-1128.
55. Büchler M, König J, Brom R et al. cDNA cloning of the hepatocyte canalicular isoform of the multidrug resistance protein, cMrp, reveals a novel conjugate export pump deficient in hyperbilirubinemic mutant rats. J Biol Chem 1996; 271: 15091-15098.
56. Goodfellow HR, Sardini A, Ruetz S et al. Protein kinase C-mediated phosphorylation does not regulate drug transport by the human multidrug resistance P-glycoprotein. J Biol Chem 1996; 271: 13668-13674.
57. Germann UA, Chambers TC, Ambudkar SV et al. Characterization of phosphorylation-defective mutants of human P-glycoprotein expressed in mammalian cells. J Biol Chem 1996; 271: 1708-1716.
58. Shaulsky G, Kuspa A Loomis WF. A multidrug resistance transporter/serine protease gene is required for prestalk specialization in *Dyctiostelium*. Genes Dev 1995; 9: 1111-1122.
59. Chen C, Ching J, Ueda K et al. Internal Duplication and Homology with Bacterial Transport Protein in the *mdr1* (P-Glycoprotein) Gene from Multidrug Resistant Human Cells. Cell 1986; 47: 381-389.
60. Cole SPC, Bhardwaj G, Gerlach JH et al. Overexpression of a transporter gene in a multidrug-resistant human lung cancer cell line. Science 1992; 258: 1650-1654.
61. Jedlitschky G, Leier I, Buchholz U et al. ATP-dependent transport of

glutathione *S*-conjugates by the multidrug resistance-associated protein. Cancer Res 1994; 54: 4833-4836.

62. O'Brien ML, Tew KD. Glutathione and related enzymes in multidrug resistance. Eur J Cancer [A] 1996; 32A: 967-978.

63. Twentyman PR, Versantvoort CHM. Experimental modulation of MRP (multidrug resistance-associated protein) mediated resistance. Eur J Cancer [A] 1996; 32A: 1002-1009.

64. Kessel D, Botteril V, Wodinsky I. Uptake and retention of daunomycin by mouse leukemic cells as factors in drug response. Cancer Res 1968; 28: 938-941.

65. Biedler JL, Riehm H. Cellular resistance to actinomycin D in Chinese hamster cells in vitro: cross resistance, radioautographic, and cytogenetic studies. Cancer Res 1970; 30: 1174-1184.

66. van der Bliek AM, Baas F, Ten Houte de Lange T et al. The human *mdr3* encodes a novel P-glycoprotein homologue and gives rise to alternatively spliced mRNAs in liver. EMBO J 1987; 6: 3325-3331.

67. Hsu SI, Lothstein L, Horwitz SB. Differential overexpression of three *mdr* gene family members in multidrug-resistant J774.2 mouse cells. J Biol Chem 1989; 264: 12053-12062.

68. Devault A, Gros P. Two members of the mouse *mdr* gene family confer multidrug resistance with overlapping but distinct specificities. Mol Cell Biol 1990; 10: 1652-1663.

69. Gros P, Croop J, Housman DE. Mammalian multidrug resistance gene: complete cDNA sequence indicates strong homology to bacterial transport proteins. Cell 1986; 47: 371-380.

70. Gros P, Raymond M, Bell J et al. Cloning and characterizationof a second member of the mouse *mdr* gene family. Mol Cell Biol 1988; 8: 2770-2778.

71. Gottesman MM, Pastan I. Biochemistry of multidrug resistance mediated by the multidrug transporter. Ann Rev Biochem 1992; 62: 385-428.

72. Borst P, Schinkel AH. What have we learnt thus far from mice with disrupted P-glycoprotein genes? Eur J Cancer [A] 1996; 32A: 985-990.

73. Schinkel AH, Smit JJM, van Tellingen O et al. Disruption of the mouse *mdr1a* P-glycoprotein gene leads to a deficiency in the blood-brain barrier and to increased sensitivity to drugs. Cell 1994; 77: 491-502.

74. Young PR, Krasney PA. Stimulation of interleukine-1β secretion in monkey kidney cells by coexpression of the mammalian P-glycoprotein *MDR1*. In: eds Ghezzi, P Mantovani, A Biomedical press, 1992; p. 21-27.

75. Ruetz S, Brault M, Kast C et al. Functional expression of the multidrug resistance-associated protein in the yeast *Saccharomyces cerevisiae*. J Biol Chem 1996; 271: 4154-4160.

76. Smit JJM, Schinkel AH, Oude-Elferink RPJ et al. Homozygous disruption of the murine *mdr2* P-glycoprotein gene leads to a complete absence of phospholipid from bile and to liver disease. Cell 1993; 75: 451-462.

77. Ruetz S, Gros P Phosphatidylcholine translocase: A physiological role for the *mdr2* gene. Cell 1994; 77: 1071-1081.

78. Ruetz S, Raymond M, Gros, P. Functional expression of P-glycoprotein encoded by the mouse *mdr3* gene in yeast cells. Proc Natl Acad Sci USA 1993; 90: 11588-11592.

79. Klugbauer N, Hofman F. Primary structure of a novel ABC transporter with a chromosomal localization in the band encoding the multidrug resistance associated protein. FEBS Lett 1996; 391: 61-65.

80 Connors, TD, Van Raay TJ, Petry Lr, Klinger KW, Landes GM, Burn TC. THe cloning of a human ABC gene (ABC 3) mapping to chromosome 16p73.3. Genomics 1997; 39:231-36.

81. Luciani MF, Chimini G. The ATP binding cassette transporter ABC1. is required for the engulfment of corpses generated by apoptoic cell death. EMBO J 1996; 15: 226-235.

82. Saunders JW, Gasseling MT. Cellular death in morphogenesis of the avian wing. Dev Biol 1962; 5: 147-178.

83. Austyn JM, Gordon S. F4/80, a monoclonal antibody directed specifically against the mouse macrophage. Eur J Immunol. 1981; 11: 805-815.

84. Surh CD, Sprent J. T cell apoptosis detected in situ during positive and negative selection in the thymus. Nature 1994; 372: 100-103.

85. Ellis RE, Jacobson DM, Horvitz HR. Genes required for the engulfment of cell corpses during programmed cell death in *Caenorhabditis elegans*. Genetics 1991; 129: 79-94.

86. Wilson R, Ainscough R, Anderson K et al. 2.2 Mb of contiguous nucleotide sequence from chromosome III of *C. elegans*. Nature 1994; 368: 32-38.

87. Ellis RE, Yuan J, Horvitz HR. Mechanisms and functions of cell death. Ann Rev Cell Biol 1991; 7: 663-698.

88. Hengartner MO, Horvitz HR. Programmed cell death in *Caenorhabditis elegans*. Curr Opin Genet Develop 1994; 4: 581-586.

89. Yuan J, Shaham S, Ledoux S et al. The *C. elegans* cell death gene *ced-3* encodes a protein similar to mammalian interleukin-1β-converting enzyme. Cell 1993; 75: 641-652.

90. Hengartner MO, Horvitz HR. *C.elegans* cell survival gene *ced-9* encodes a functional homolog of the mammalian proto-oncogene *bcl-2*. Cell 1994; 76: 665-676.

91. Hengartner MO, Horvitz HR *C.elegans* gene *ced-9* protects cells from programmed cell death. Nature 1992; 356: 494-499.

92. Henkart PA. ICE Family Proteases: Mediators of all apoptotic cell death? Immunity 1996; 4: 195-201.

93. Becq F, Hamon Y, Bajetto A et al. ABC1, an ATP binding cassetta transporter required during apoptosis, generates a regulated anion flux after expression in *Xenopus* oocytes. J Biol Chem 1996; (In Press)

94. Dinarello, CA. Biological basis for interleukin-1 in disease. Blood 87: 2095-2147. 1996;

95. Perregaux DG, Gabel CA. IL-1β maturation and release in response to ATP and nigericin . J Biol Chem 1994; 269: 15195-15200.
96. McNiff P, Svensson L, Pazoles CJ et al. Tenidap modulates cytoplasmic pH and inhibits anion transport in vitro I. mechanisms and evidence of functional significance. J Immunol 1994; 153: 2180-2193.
97. Laliberte R, Perregaux D, Pazoles CJ et al. Tenidap modulates cytoplasmic pH and inhibits anion transport in vitro. II Inhibition of IL-1β production from ATP-treated monocytes and macrophages. J Immunol 1994; 153: 2168-2179.
98. Perregaux DG, Svensson L, Gabel CA. Tenidap and other anion inhibitors disrupt cytosolic T lymphocyte-mediated IL-1β post-translational processing. J Immunol 1996; 157: 57-64.
99. Cerreti DP, Kozlosky CJ, Mosley B et al. Molecular cloning of the interleukin-1β converting enzyme. Science 1992; 256: 97-100.
100. Thornberry NA, Bull HG, Calaycay JR et al. A novel heterodimeric cysteine protease is required for interleukin-1β processing in monocytes. Nature 1992; 356: 768-774.
101. Ayala JM, Yamin TT, Egger LA et al. IL-1β-converting enzyme is present in monocytic cells as an inactive 45 kDa precursor. J Immunol 1994; 153: 2592-2599.
102. Singer II, Scott S, Chin J et al. The interleukin-1β-converting enzyme (ICE) is localized on the external cell surface membranes and in the cytoplasmic ground substance of the human monocytes by immunoelectron microscopy. J Exp Med 1995; 182: 1447-1459.

CHAPTER 6

SECRETORY LYSOSOMES AND THE PRODUCTION OF EXOSOMES

Graça Raposo, Michel Vidal and Hans Geuze

I. INTRODUCTION

GENERAL OUTLOOK

Eukaryotic cells secrete proteins either by the so-called constitutive secretion involving vesicular transport and exocytosis or by the regulated secretion of storage granules upon proper stimulation.[1,2] Only recently has one become aware that alternative mechanisms operate that may account for the secretion of specific membrane and cytosolic proteins. Indeed, accumulating evidence indicates that cellular compartments displaying intralumenal membrane vesicles, collectively named multivesicular bodies (MVBs), fuse with the plasma membrane in an exocytic manner. During exocytosis, the 60 to 80 nm membrane vesicles present in the lumen of MVBs are released into the extracellular environment. The secreted membrane vesicles are called exosomes.[3-6]

On the other hand, MVBs have for a long time been known as subcompartments of the endocytic route and constitute the late endosomal or prelysosomal system of the eukaryotic cell.[7-9] Moreover, they also represent a meeting point between the endocytic and the exocytic pathways, since particular molecules access these compartments after egress from the trans-Golgi network (TGN), either directly by vesicular transport, or indirectly via early endosomes.[10,11] Nevertheless, at present the definition of the late endosomal, prelysosomal MVB should be reconsidered and extended as data obtained by several laboratories indicate that secretory granules, particularly in cells of hematopoietic origin, are closely related to lysosomes. They are therefore called secretory lysosomes.[12] The cytotoxic granules of cytotoxic T lymphocytes (CTLs) and Natural Killer (NK) cells, the secretory granules of mast cells and basophils, and the α-granules in blood platelets share several characteristics such as

Unusual Secretory Pathways: From Bacteria to Man, edited by Karl Kuchler, Anna Rubartelli and Barry Holland. © 1997 R.G. Landes Company.

the presence of intraluminal 60-80 nm membrane vesicles, their late accessibility to endocytic tracers and the expression of lysosomal marker proteins[13-16] (Heijnen and Geuze, unpublished data on platelets; Raposo, Bonnerot and Desaymard, unpublished data on mast cells). Furthermore, illustrating the relationship between prelysosomal MVBs and secretory granules, the lysosomal disorder Chediak-Higashi syndrome affects also secretory granules of mast cells, platelets, neutrophils and cytotoxic T cells.[17-19] However, it has not yet been established how secretory lysosomes are formed and whether they represent a sub-population of lysosomes.[14,15] The fact that cells that exhibit secretory lysosomes are all derived from the hematopoietic lineage suggests that these cells may possess specialized secretory mechanisms that allow lysosomes to be secreted.[12]

The capacity of MVBs to fuse with the cell surface in certain cell types and release their contents including exosomes into the extracellular environment shows that these (pre)lysosomal compartments are not merely dead end points in the endocytic/degradative pathway. Despite accumulating evidence indicating the potential role of the so-called secretory lysosomes,[12] the physiological relevance of exosome release is still equivocal and awaits further research. Data obtained by several laboratories have shown that exosomes released by reticulocytes are responsible for the eradication of transferrin receptors (TfRs) during their maturation to erythrocytes, a crucial step in the differentiation process.[3,20-22] Major histocompatibility complex (MHC) class II molecules that are present on exosomes released by B lymphocytes are able to stimulate T cells in an antigen-dependent manner.[6] Exosomes released by cytotoxic T cells and NK cells upon their interaction with target cells contain perforin and granzymes and have the ability to generate a cytotoxic response.[23,24]

In the first section of this chapter we summarize the current concepts of intracellular traffic in eukaryotic cells and, in particular, we describe the cellular compartments playing key roles in endocytosis and exocytosis. In the second section we will focus on recent research on two cellular systems, reticulocytes and B lymphocytes, for which an alternative pathway of secretion via exosomes has been characterized. In this context, we summarize data indicating a role of small membrane vesicles secreted by cytotoxic and NK cells. Finally, since MVBs are present in almost all eukaryotic cells, exosome release could operate as a generalized alternative route for secretion of membrane and cytosolic proteins.

Intracellular compartmentalization: An overview

The endocytic route: Endosomal and lysosomal compartments
Eukaryotic cells use the endocytic pathway for the uptake of nutrients and growth factors via specific membrane receptors. The process of receptor-

mediated endocytosis starts with binding of a ligand to its receptor on the plasma membrane and internalization of the complex via clathrin-coated pits and vesicles.[9,25] After losing their coat, vesicles with their cargo fuse with early endosomes (Fig. 6.1). Therefore, in the endocytic pathway early endosomes are the first intracellular compartment reached by internalized molecules. At the ultrastructural level they appear as a network of tubules and irregularly shaped vacuoles often surrounded by buds and vesicles.[26-31] The early endosomal compartment is considered a sorting and recycling compartment, since it is able to segregate molecules that must be recycled to the plasma membrane from molecules destined to be degraded. Studies on a number of receptor species such as the low density lipoprotein receptor (LDLR), the asialoglycoprotein receptor (ASGPR), or the transferrin receptor (TfR) allowed the definition of sorting events occurring at the early endosome level.[32,36,37] Recycling from early

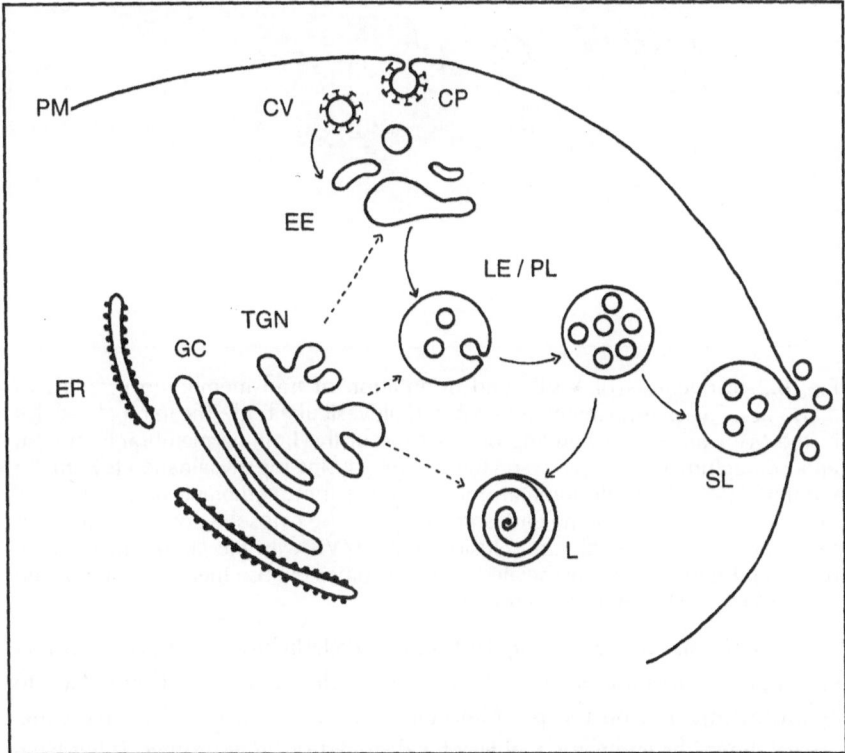

Fig. 6.1. Schematic representation of the endocytic and exocytic system of eukaryotic cells. PM: plasma membrane; GC: Golgi complex; ER: endoplasmic reticulum. CP: coated pit ; CV: coated vesicles; EE: early endosomal compartment; LE/PL: late endosomal/prelysosomal compartment (MVBs); L: lysosomal compartment; SL: secretory lysosome

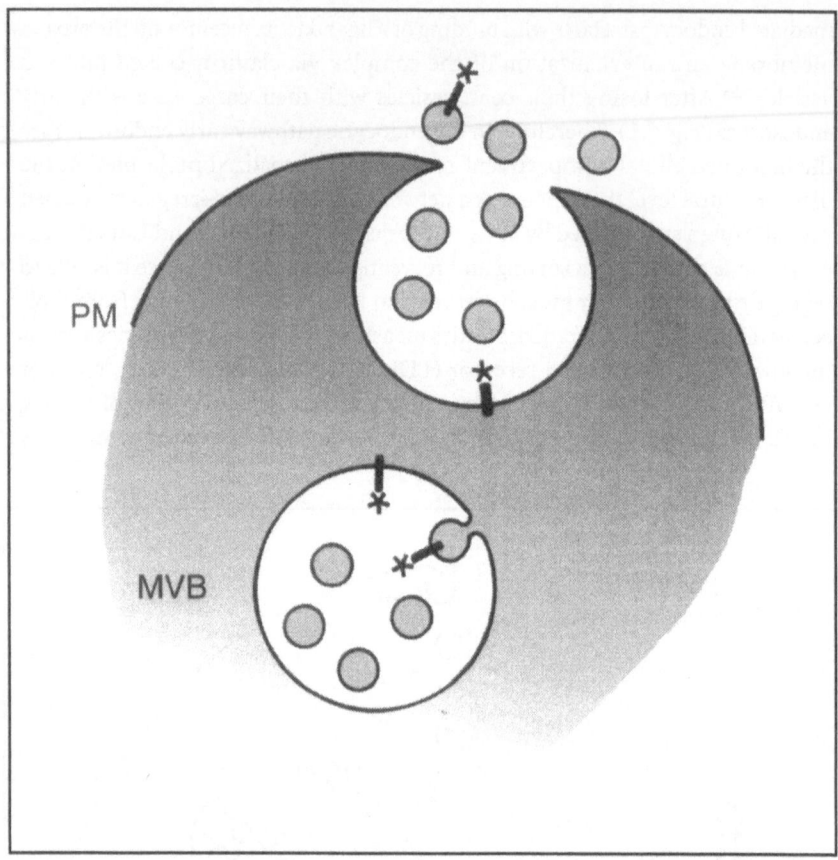

Fig. 6.2. Biogenesis of MVBs and orientation of transmembrane proteins in exosomes. PM: plasma membrane; MVB: multivesicular body. The internal vesicles of the MVB arise from budding of a portion of the limiting membrane into the endosomal lumen. As depicted on the scheme, during the invagination toward the intralumenal milieu, the luminal domain of the transmembrane protein (star) is oriented to the lumen of the endosome whereas the cytosolic domain remains in the intravesicular cytosol. Upon fusion of the MVB with the plasma membrane, transmembrane proteins associated with exosomes expose their luminal domain (star) into the extracellular environment.

endosomes may occur by vesicular traffic, and clathrin-coated vesicles distinct from plasma membrane, or TGN derived vesicles are potential candidates to perform this transport step.[38] Molecules that are not destined for recycling, travel down the endocytic pathway for degradation in lysosomes. It has been proposed that endocytosed molecules may be transported from early endosomes to pre-existing late endosomes and lysosomes by vesicular transport.[39] However, evidence has accumulated in favor of a model in which early endosomes gradually mature into late endosomes and lysosomes.[40-42] Alternatively, the maturing endosome may fuse with pre-existing lysosomes.[43] At this point one must consider the ultrastructural features and the biogenesis of the late endosomal/prelysosomal system. At the electron microscopical level the late endosome, or prelysosome, appears as a 200 to 300 nm membrane compartment containing in its lumen variable amounts of small 60-80 nm vesicles (Fig. 6.1). Because of this feature it is commonly named the multivesicular body (MVB).[7,8,31] The internal vesicles of the MVB are thought to arise from budding of a portion of the limiting membrane into the endosomal lumen.[8,9] During the invagination toward the intralumenal milieu some membrane proteins are sequestered in the internal vesicles, whereas others remain in the limiting membrane of the MVB. During formation of the internal vesicles small buds of cytosol are trapped within the vesicle (Figs. 6.1 and 6.2). The best example of protein sorting within the MVB comes from studies on the intracellular processing of the Epidermal Growth Factor Receptor (EGFR). Whereas recycling TfRs remain in the limiting membrane of the maturing MVB, EGFRs destined for degradation in lysosomes, are sorted into the internal vesicles.[44,45] This reveals an attractive bifunctional model in which the early endosome matures into a late multivesicular endosome, while continuously recycling membrane proteins to the plasma membrane by vesicular transport and sequestering others into intralumenal membranes. The subcellular mechanisms implicated in such sorting events remain poorly understood, though it has been shown that the sequestration of the EGFR into the internal vesicles requires kinase activity.[45] The most prominently phosphorylated substrate of the EGFR kinase is annexin 1, a protein that interacts directly with both phospholipids and actin.[46] Our unpublished data suggest that clustering of proteins and lipids must precede sequestering into the internal vesicles, implying that chaperone proteins may be involved in the specific clustering of membrane components (Vidal et al, submitted). As detailed in the next section, a candidate chaperone is the cognate heat shock protein Hhsc70 which has been identified as an uncoating ATPase.[47,48] Recently, a cytosolic protein from the nexin family, named Snx-1, has been shown to interact directly with the cytoplasmic domain of the EGFR.[49] This protein and probably other members of this newly identified family could play a role in the sorting events associated with the MVB.

Just as in the case of endosome formation, the biogenesis of lysosomes and the question of how molecules are transferred to lysosomes is still poorly understood. On the basis of their protein content, morphology and acidity, lysosomes seem distinct from late endosomes. While late endosomes are enriched in the mannose-6-phosphate receptor (MPR), whose main function is to transport newly synthesized lysosomal enzymes from the Golgi complex, lysosomes are mostly devoid of MPR and have a lower pH.[10,50,51] Also, lysosomes appear in the electron microscope as denser organelles with internal membrane sheets and amorphous material. However, hydrolytic enzymes (cathepsins, β-hexosaminidase) and lysosomal membrane proteins (Lamps, Limps, LAP) can also be detected in the late endosomal/prelysosomal compartment. Thus, lysosomes may represent the final stage of maturation of the late endosome, because the latter loses endosomal constituents and gradually acquires proteins characteristic of lysosome.[8,40,42]

The exocytic route: Constitutive and regulated secretion, secretory granules and secretory lysosomes

Secretory proteins are unable to diffuse through the plasma membrane; they cross lipid bilayers only during synthesis when they are transported through the membrane of the rough endoplasmic reticulum (RER). From the RER they are transferred by vesicular transport to the Golgi complex where they can be processed and terminally glycosylated before being sorted in the trans-Golgi network (TGN) for transport to the plasma membrane. Two types of transport can be discerned. First, constitutive secretion is mediated by small vesicles. These vesicles have not been characterized as yet, and other structures have not been revealed. Specialized secretory cells such as exocrine and endocrine cells and certain neurons are able to store newly synthesized proteins in larger secretory granules that are exocytosed upon external stimuli.[2]

Little is known in nonpolarized cells about the cellular and molecular machineries that allow newly synthesized secretory and plasma membrane proteins to be packaged in either form of secretory vesicle from the TGN to the cell surface. Surface delivery by constitutive secretion seems to occur as a bulk flow transport without the requirement of sorting signals for cargo proteins.[52] On the other hand, in polarized epithelial cells the molecular machineries implicated in apical and basolateral sorting of proteins have been the topic of intensive studies.[53,54] Some evidence suggests that both the apical and basolateral cognate routes may operate in nonpolarized cells,[55] but the vesicular intermediates involved in constitutive secretion are still poorly characterized.

More is known on the subcellular mechanisms operating in specialized secretory cells. The biogenesis of secretory granules likely occurs at the TGN by proteins condensating into soluble aggregates.[56,57] The selective aggregation of secretory proteins in the TGN is thought to be triggered by the lumenal milieu of the compartment, especially by the weakly acidic pH and a high calcium concentration.[58] Sorting receptors may direct soluble protein aggregates from the TGN to secretory granules.[59] However, in an alternative model secretory proteins are only sorted after reaching the secretory granule, the so-called sorting by retention model.[60]

Electron microscopical observations of hematopoietic cells have shown that secretory granules can show heterogeneity in their content and, as detailed below, contain proteins that are also found in (pre)lysosomes. Three types of distinct granules can be distinguished morphologically in CTLs,[13] NK cells[15] and mast cells.[16] Type I granules have an homogeneous electron-dense core, whereas Type II granules are more irregularly shaped and contain small membrane vesicles and multilaminar membranes. Intermediate granules (Type III) display an electron-dense core surrounded by a large multivesicular cortex. The multivesicular portion of Type II and Type III granules is reminiscent of the multivesicular prelysosome described above. In agreement with the lysosomal nature of the granule, they are highly enriched in lysosomal membrane proteins (Lamp 1, Lamp 2, CD63) and lysosomal enzymes (cathepsins, β-hexosaminidase). In contrast to mature lysosomes and emphasizing their similarity to a late endosomal/prelysosomal compartment, they may contain the mannose-6-phosphate receptor (M_6PR).[10,15] Secretory granules in CTLs can also accumulate endocytosed proteins illustrating their close connection to the endocytic pathway.[13,15] In contrast to lysosomal markers and proteoglycans which represent the major constituents of the dense core in CTLs granules, other components differ depending on the cell type. For example, in mast cells the granules are enriched in histamine and serotonine, whereas in cytotoxic T cells and NK cells they are enriched in granzymes and perforin. After specific triggering, for example by interaction with a target cell, the three types of granules are able to fuse with the plasma membrane. Concomitantly with soluble molecules, membrane proteins present in the internal vesicles of Type II and Type III granules are secreted. Thus, in these specialized secretory cells no clear distinction can be made between secretory granules and lysosomes, but rather they contain dual-function organelles, the secretory lysosomes. These secretory lysosomes must ensure both a main cellular function such as target cell destruction by cytotoxic T cells and NK cells and maintenance of the normal turnover of its molecules.[15]

Several other cells secrete their lysosomal content to ensure their main functions. Macrophages and neutrophils defend tissue homeostasis and participate in inflammatory responses by secreting oxidative metabolites, cytolytic proteins, or degradative enzymes contained in lysosomal compartments.[12,61] In the lung, the surfactant proteins synthesized by alveolar Type II cells are secreted during fusion of lysosomal multilamellar bodies with the plasma membrane.[62] The osteoclast is another example of a cell of hematopoietic origin possessing secretory lysosomes whose main function is to ensure bone resorption.[63]

Based on evidence accumulated by several observations in agreement with the concept of secretory lysosomes, the assumption that lysosomes do not always represent the dead end stages of cell constituents in the endocytic route is reevaluated. In the next section we will extend this notion by detailing studies of two cell systems in which the contents of secretory (pre)lysosomal compartments may have important functional implications in cell physiology.

II. CURRENT RESEARCH

Secretion of Exosomes by Hematopoietic Cells

As described in the introductory section several cells contain secretory lysosomes from which the contents are released in a regulated manner. The main characteristic of the secretory lysosome is the presence of a dense core mainly composed of proteases and proteoglycans. In addition, electron microscopical studies have revealed in their lumen membrane vesicles with a diameter ranging from 60 to 80 nm and in some cases membrane sheets.[14,15,64] As already mentioned, the molecular mechanisms allowing the formation of the intralumenal vesicles of the prelysosome, their membrane and cytosolic components, as well as their putative role upon secretion are not yet clarified. Recent research on the intracellular trafficking of transferrin receptors (TfRs) in reticulocytes and MHC class II molecules in B cells has further defined the exocytic process of MVBs and has revealed the characterization of the membrane vesicles, exosomes, released during exocytosis.

Reticulocytes

Function and intracellular transport of the transferrin receptor

Iron is a critical component for biochemical reactions, but it is also essential for heme formation in developing erythroid cells. Because of the instability of Fe^{3+} ions in biological fluids, nature has developed efficient iron-carrier proteins, the most important being the plasma protein transferrin. Iron is bound to the carrier transferrin at neutral pH ($Ka \approx 1\text{-}6 \times 10^{22}$ M^{-1}), but dissociation gradually increases as pH decreases. Moreover, transferrin interaction with the transferrin receptor largely depends on the degree of iron saturation of transferrin and the local pH value. The cell exploits the acid sensitivity of both the iron-transferrin bond and the ligand-receptor interaction in the iron accumulation process.

Transferrin receptors are present on the cell surface of a wide range of eukaryotic species and cell types, where they can bind transferrin. The ligand-receptor complex enters the cell by clathrin-coated pits, and after clathrin uncoating by a specific ATPase is delivered to early endosomes during fusion events.[65] Progression of internalized molecules along the endocytic pathway is associated with a more acidic environment ranging from pH 6.5 down to pH <5 in the lysosomes.[66] In the acidified endosomal compartment, iron is dissociated from transferrin and translocated across the endosomal membrane to reach the cytosol. Using the ^{+}H-ATPase from reticulocyte endosomes reconstituted in liposomes, a direct role has been recently suggested for the vacuolar pump in iron transport.[67] At the same time, apotransferrin remains attached to its receptor and is recycled to the plasma membrane. Exposed to the neutral pH of the medium, the ligand dissociates from the receptor which is free to bind another transferrin molecule.[68] As specified in the introductory section, other ligands and receptors reaching the same early endocytic compartments are targeted to late endosomes and lysosomes.[9] Despite the fact that in most cells the TfR uses the recycling early endosome to exert its main function, in reticulocytes, as detailed below, the receptor is targeted to MVBs.[3,20]

Secretion of transferrin receptor-enriched exosomes

During its differentiation into an erythrocyte, the reticulocyte loses all its internal compartments: endosomal compartments, mitochondria, remnants of ER and Golgi apparatus and also several membrane-associated activities. For example, mammalian red cells lose all their TfR during maturation.[69] This loss of Tf binding activity was first thought to be due to inactivation of the receptors by proteolysis. It is now accepted to occur by selective removal of membrane proteins by vesiculation.[3,5,20] The functional proteins are released into the extracellular medium following initial internalization and packaging in 60-80 nm membrane vesicles in the lumen of late endosomes.[21] Indeed, thin sections of maturing reticulocytes showed the presence of intracellular MVBs.[3,20] The vesicles in the lumen of these MVBs had the same size of those released into the extracellular medium during reticulocyte maturation. Moreover, the concomitant presence of these extracellular vesicles, the so-called exosomes, and multivesicular organelles in the cells has been observed during both in vitro and in vivo experiments, suggesting a physiological process.

Exosomes carrying TfR can be collected from the extracellular medium by differential ultracentrifugation steps and analyzed for their composition. It has been shown that the lipid composition and asymmetry are similar to those of the plasma membrane.[5] The protein composition, however, is very different between exosomes and plasma membrane.[5,21] Especially, two major exosomal proteins were identified as the TfR and the clathrin-uncoating ATPase (UC-ATPase),[48] whereas major plasma membrane spanning proteins (e.g., band 3) or cytosolic enzymes (e.g., lactate dehydrogenase) were not detected in exosomes. Other membrane-associated activities are found in exosomes;[21] all are known to diminish from the plasma membrane during reticulocyte maturation. Apart from the plasma membrane activities lost during reticulocyte maturation, lysosomal activities (e.g., N-acetyl β-glucosaminidase) were also found in exosomes.[70]

As referred to in the introductory section, the signals responsible for the packaging of proteins in exosomes are not known. Also, the presence in the lumen of exosomes of the UC-ATPase, a cytosolic protein, suggested that this protein may be involved in selecting proteins destined to be sequestered in intralumenal membrane vesicles of the MVB.[48] This protein is the cognate form of the 70 kDa heat shock protein and possesses chaperone characteristics. It may bind to unfolded or partly denatured proteins, such as the cytosolic domain of the Tf receptor and contribute to its sorting to exosomal membranes during the formation of the MVB.[48] This protein may either hide a domain involved in regulation of the recycling step[71] or interact with the sorting machinery.[72] All these events may represent a "quality control" event in the endocytic pathway similar to the one occurring during protein synthesis in the ER. However, other mechanisms may allow segregation of proteins in exosomes. Acetylcholine esterase, for example, has been shown to be released in exosomes with concentrations compatible with a 50% decrease from the cell surface during red cell maturation.[21] In agreement, we have recently shown that in opposition to C6-NBD-SM (N-(N-[6-[(7-nitrobenz-2-oxa-1,3-diazol-4-yl)amino]-caproyl])sphingomyelin) which efficiently recycles to the plasma membrane, N-Rh-PE (N-(lissamine rhodamine B sulfonyl-phosphatidylethanolamine) is actively sorted after its internalization by reticulocytes, and packaged in exosomes (Vidal et al, unpublished data). Thus, as for acetylcholine esterase, N-Rh-PE located in the exoplasmic leaflet due to experimental conditions is retrieved from other phospholipids and released in the extracellular medium. In the case of the fluorescent phospholipid analog, self-aggregation in endosomes may be the signal triggering its sorting to exosomes. Indeed, we showed that aggregation of TfR on the exoplasmic side of the plasma membrane induced by specific antibodies increases the release of the receptors in exosomes (Vidal et al, unpublished data). It is worth noting that the property of self-aggregation of glycosphingolipids has been suggested to be involved in sorting steps both during synthesis, at the TGN level[73] and degradation in multivesicular structures.[74]

The main function of exosome release during reticulocyte maturation is thought to be the clearing of several obsolete proteins. The TfR is the major protein which completely disappears from the red cell plasma membrane during its differentiation into an erythrocyte. The red cell does not need to uptake iron anymore and since it may generate free radicals, iron is poisonous for a cell that has lost its capacity to replace damaged proteins. Thus, to overcome this danger the TfR, like the EGF-R,[75] must undergo a downregulation event in the MVB. The release of TfR-containing exosomes in blood has allowed the development of assays quantifying the level of the circulating receptor, and gives indications of the degree of mild iron deficiency.[76] However, the fate of exosomes in the circulation is not known. However, since exosomes lack aminophospholipid translocase activity,[5] phosphatidylserine may be exposed on the exoleaflet of the exosome membrane—a feature that should help their uptake and processing by macrophages.

B Lymphocytes

Function and intracellular transport of MHC class II molecules

T cell proliferation and T cell cytotoxicity are initiated following the specific recognition by the T cell receptor of a tight complex of MHC class II, or class I molecules and antigenic peptides expressed on the surface of antigen presenting cells (APCs). MHC molecules are viewed as polymorphic membrane receptors able to bind specifically a peptide ligand with a particular length and sequence. A major breakthrough toward the understanding of where MHC molecules meet antigenic peptides and how these are presented to T cells has been accomplished in the last six years by a combination of biochemical and morphological approaches similar to those conceived to resolve the intracellular transport of other membrane receptors.[77,78]

In this section we summarize the current concepts of the intracellular traffic of MHC class II molecules in APCs in connection to the processing, binding and presentation of antigens. Substantial recent evidence has shown the intimate relationship between the endocytic and the secretory pathway, and has revealed the idea that an alternative pathway of secretion involving exosomes may operate in APCs.

MHC class II molecules expressed in APCs such as B cells, macrophages and dendritic cells, are composed of two transmembrane polypeptides, the α and the β chain, with a MW of 32-35 kDa and 28-30 kDa, respectively. After their biosynthesis in the endoplasmic reticulum (ER) the α,β chains associate with a third polypeptide, the invariant chain (I-chain), which is a Type II membrane protein.[79] A trimer of I-chain binds three MHC class II $\alpha\beta$ dimers, whereby a particular sequence in the luminal C terminal domain of the I-chain occupies the binding groove of MHC class II, thereby preventing endogeneous peptide-binding in the ER. This sequence includes residues 81-104 of the I-chain, and is designated "MHC class II associated invariant chain peptides" (CLIP).[80] The MHC class II/ I-chain complexes are glycosylated in the Golgi complex and at the TGN side, they are segregated from the constitutive secretory route followed by MHC class I molecules.[81,82] These complexes are targeted to the endocytic system where the I-chain is degraded by proteases.[83,84] The removal of CLIP is a critical step for the peptide presenting capacity of MHC class II. The recently identified and characterized class II-like molecule, HLA-DM, fulfills a catalyzing role in the removal of CLIP which renders MHC class II free to bind antigenic peptides.[85,86] These are of a particular length and sequence and arise from the degradation of antigens taken up by the APC via different routes, including fluid phase and receptor-mediated endocytosis and phagocytosis. Mature MHC class II molecules, with tightly bound antigenic peptides, must then be transported from the endocytic pathway to the cell surface to accomplish their antigen presentation function.[77,87]

Although many details of the MHC class II transport pathway have recently been resolved, other important questions remain to be answered. Thus, only little is known about the critical step of peptide-loaded MHC class II molecules transport to the plasma membrane. It is unknown by which vesicular carriers this transport is mediated, from where in the endocytic system such vesicles originate and how this transport is regulated.

Using immunocytochemistry and the electron microscope, Peters and collaborators have shown that in B lymphoblastoid cells, the majority of the intracellular MHC class II molecules is localized in lysosome-related compartments, designated MHC class II compartments (MIICs).[82] These compartments display concentrically arranged multilaminar membranes, are mildly acidic and contain lysosomal enzymes (β-hexosaminidase, cathepsin D) and lysosomal membrane proteins (LAMPs, CD63). Further investigation of MHC class II localization by immunogold cytochemistry showed that in B cells, macrophages and dendritic cells, including the Langerhans cells of the skin, MIICs are a more heterogeneous population of structures than originally thought.[6,88-90] At least four types of MIICs can be discerned; multilaminar, multivesicular: intermediate types and MIICs with only a few internal vesicles and an irregular shape (Fig. 6.3).[91] All subtypes of MIICs contain lysosomal components, though they accumulate different amounts of invariant chain and show different accessibility to endocytic tracers.[90,91] MIICs with internal vesicles, reminiscent of MVBs in other cell types, are positioned earlier in the endocytic pathway than those containing membrane sheets.[6,90,91] Biochemical studies using subcellular fractionation as well as functional assays were carried out by several laboratories to analyze the direct implication of MIICs in antigen presentation.[92-95] These studies; together with the observations that HLA-DM[89,96] and class II-peptide complexes (Morkowsky, Raposo, Geuze and Rudensky; Jm Press E.J Immunol.) accumulate in MIICs, suggest that the acidic and protease-rich MIICs may indeed represent the meeting point between MHC class II molecules delivered from the biosynthetic pathway and antigenic peptides derived from endocytosed proteins. Studies on splenic cells and on the murine B cell line A20 open the possibility that depending on the APC and most likely on the antigens to be handled, compartments positioned earlier in the endocytic pathway, similar to early endosomes, may be involved in peptide binding as well.[97,98]

How are MHC class II molecules transferred to the cell surface? Lysosomes are thought to be the dead-end stages of the endocytic route. Little is known about transport routes allowing egress from lysosomes or any other endocytic compartment to the cell surface. Lysosomal membrane proteins like LEP100 or Lamp 1 have been detected on the cell surface suggesting transport to the cell

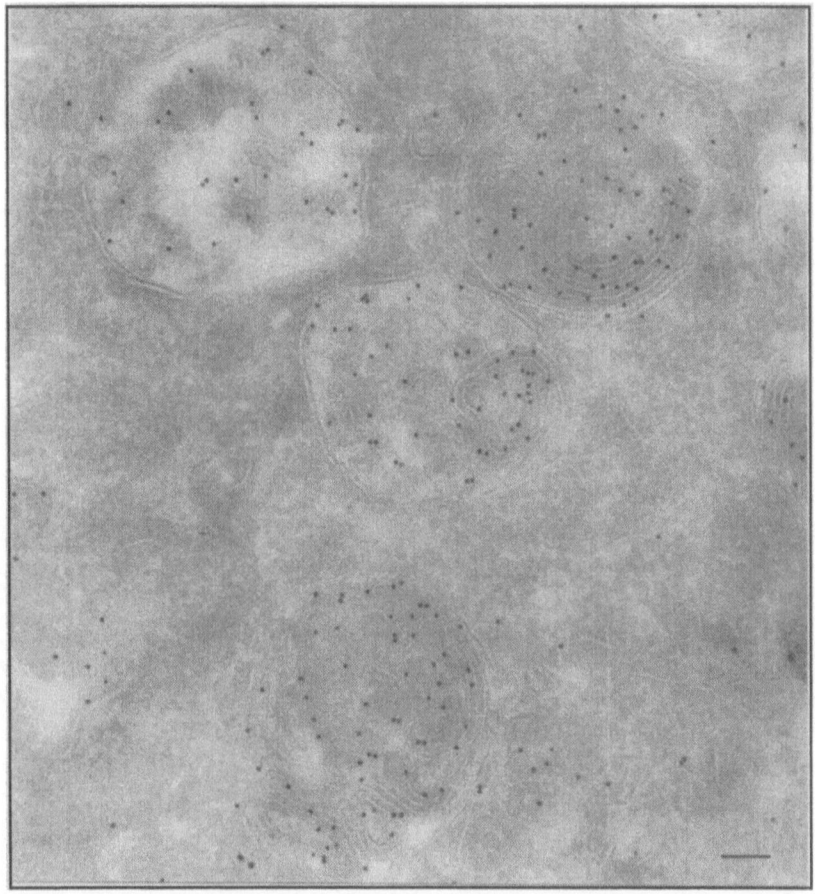

Fig. 6.3. MHC class II compartments (MIICs) of human B lymphocytes. Ultrathin cryosections of B-lymphoblastoid cells were immunogold labeled for MHC class II (Protein A -10 nm gold). MHC class II are localized in compartments displaying internal membrane vesicles (multivesicular MIICs), membrane sheets (multilaminar MIICs) or both (intermediate MIICs). Bar: 100 nm

surface.[11,99,100] MHC class II may use a similar route for transport to the cell surface. Lysosomal components could be conveyed to the TGN and then to the cell surface by the so-called constitutive pathway. Alternatively, they may be transferred to early endosomes from which recycling to the plasma membrane may take place.

Apart from such possible transport routes of lysosomal constituents to the cell surface, our ultrastructural studies have recently revealed that multivesicular MIICs in B cells can fuse with the plasma membrane in an exocytic fashion[6] similar to that described for secretory lysosomes (Fig. 6.4). Fusion of MIICs with the plasma membrane has two major consequences: First, it provides for a direct transport of MHC class II present in the limiting membrane of the MIIC directly to the plasma membrane. Second, as a consequence of the exocytosis of MIICs, the cells release small 60-80 nm vesicles, i.e., exosomes, into the extracellular environment. These exosomes carry in their membrane MHC class II molecules with their peptide-binding site oriented toward the extracellular environment (Fig. 6.2).

Secretion of MHC class II-enriched exosomes

So far only B lymphoblastoid cell lines have been shown to secrete exosomes. Present studies in our laboratories address the possible occurrence of exosomes in physiological APCs. Recent data show that in murine bone marrow derived mast cells MHC class II localizes to secretory granules with all the characteristics of secretory lysosomes. In mast cell granules, MHC class II together with lysosomal membrane proteins are present in 60-80 nm vesicles surrounding an electron-dense core enriched in serotonine. In these cells class II-containing exosomes are only secreted upon degranulation induced by IgE-antigen complexes, indicating that exosome release is likely to be a regulated secretory process (Raposo, Bonnerot and Desaymard; unpublished data).

Exosomes represent released internal vesicles of MVBs and can be obtained from the culture media of B-lymphoblasts in relatively high quantities. This reveals the unique opportunity to analyze the composition of these vesicles and to investigate their formation. We have started to characterize exosomes by morphological and biochemical criteria. As described for the isolation of exosomes released by reticulocytes, we have used differential ultracentrifugation procedures to purify exosomes from cell culture media of different B lymphoblastoid cell lines.[6] Exosomes devoid of plasma membrane proteins, and early endosome markers, can be recovered from B cell culture supernatants after ultracentrifugation at 70,000 x g. When analyzed by immunoelectron microscopy, membranes pelleted at 70,000 x g represented an homogeneous population of vesicles that labeled for MHC class II. Western-blot analysis further showed that MHC class II associated with exosome preparations are mature SDS-stable molecules associated with peptides. Using metabolic labeling we found that within 24 hours as much as 10% of newly synthesized class II molecules are released into the extracellular media via exosomes indicating that this is not a minor pathway.

One of the major components of the exosomal membranes is MHC class II, though other still unidentified proteins are also clearly enriched.[6] Our preliminary observations indicate that exosomes may also express adhesion and coactivator molecules, suggesting that they are capable of directly interacting with T cells expressing their counterparts (Kleijmeer, Raposo and Geuze; unpublished observations). This possibility has been further tested by performing

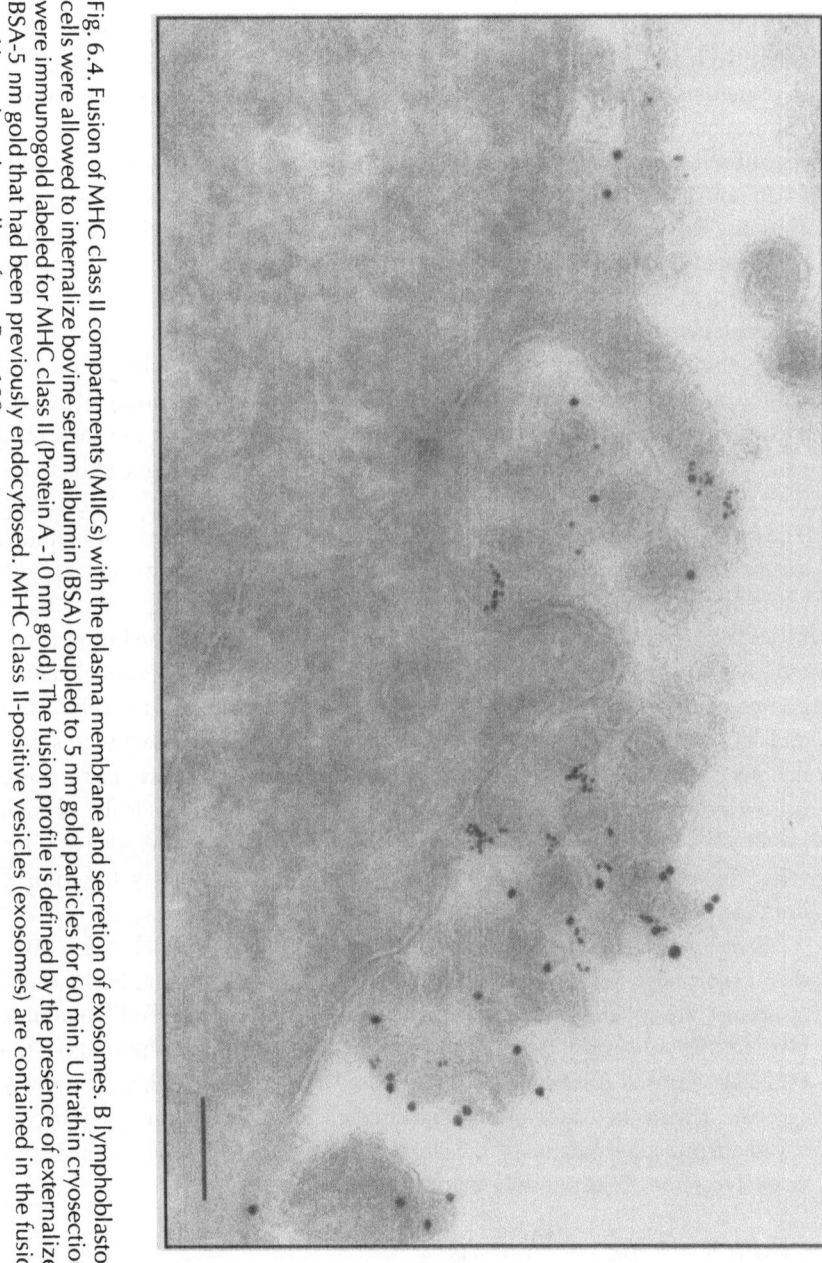

Fig. 6.4. Fusion of MHC class II compartments (MIICs) with the plasma membrane and secretion of exosomes. B lymphoblastoid cells were allowed to internalize bovine serum albumin (BSA) coupled to 5 nm gold particles for 60 min. Ultrathin cryosections were immunogold labeled for MHC class II (Protein A -10 nm gold). The fusion profile is defined by the presence of externalized BSA-5 nm gold that had been previously endocytosed. MHC class II-positive vesicles (exosomes) are contained in the fusion profile and on the cell surface. Bar: 100 nm

proliferation assays in which exosomes derived from cells primed with an anti-gen were incubated with a specific T cell clone. We have shown that exosomes are indeed able to stimulate T cell proliferation in an MHC class II haplotype-restricted manner.[6]

Whether exosomes perform a function in vivo remains to be established. For instance, exosome secretion may explain a number of observations for which the subcellular mechanisms remained unresolved. It has been reported that MHC molecules can be actively shed from murine spleen cells in short-term cultures. The release process required viable cells, was dependent on an intact cytoskeleton and the shed molecules could be recovered in the culture super-natants of splenic cells, or B cell lines.[101-103] Interestingly, in these studies MHC molecules were found to be released in the form of supramolecular particles in association with membrane lipids. The absence of other membrane proteins led the authors to interpret these results as a nonrandom shedding of plasma membrane. These observations together with the finding that the shed mol-ecules can be sedimented by ultracentrifugation at 100,000 x g lead us to be-lieve that these MHC class II containing particles actually represent exosomes. The finding that noncell associated MHC molecules are present in the mouse serum suggests that exosomes may occur in vivo.[104] Concerning their role, dif-ferent hypotheses are worth investigating. Exosomes could represent the ve-hicles transferring MHC molecules between different cells of the immune sys-tem. For example, follicular dendritic cells (FDCs) in tonsil germinal centers do not synthesize MHC class II molecules, but express them at the cell surface and act as antigen presenting cells once they have been in contact with super-natants from B cells.[105] These surface MHC class II molecules may be recruited from exosomes released by B lymphocytes. Another hypothesis is the possible implication of MHC-exosomes in the maintenance of T cell memory or toler-ance.[6] Finally, in contrast to T cell activation, exosomes may function in immunosuppression.

Exosomes released by APCs may not only carry MHC class II, but also MHC class I molecules. We have accumulated evidence indicating that MHC class I molecules are indeed present in multivesicular MIICs of several B lympho-blastoid cells (Kleijmeer, Rabouille, Raposo and Geuze; unpublished data). Thus MHC class I molecules may also be released via exosomes. This could be in agreement with the observations that shed class I molecules are similarly re-covered from supernatants of B cells.[102] Furthermore MHC class I shed from viable lymphocytes can induce cytotoxic responses.[106]

CYTOTOXIC T CELLS AND NK CELLS

As mentioned in the introductory section, cytotoxic T cells (CTLs) and NK cells display cytolytic granules with the characteristics of secretory lysosomes, since they contain lysosomal components, are acidic and are accessed by endocytic tracers.[12,14,15] CTL granule exocytosis has been considered to be an important mechanism of lethal hit delivery.[107-109] This was supported by the subcellular localization within the granule of molecules necessary for CTL-target cell interaction (TCR, CD3, CD8, MHC class I) and molecules involved in target cell killing (perforin and granzymes D, E and F).[13,24] Interestingly, within the granule the former are localized in small 60-80 nm vesicles with their luminal domain facing outward which is a consequence of the way intralumenal vesicles of prelysosomes originate, i.e., by budding from the limiting membrane. The latter accumulate in the electron-dense core, but can also be detected within the small vesicles. Upon interaction with the target cell, CTL granules fuse with the plasma membrane and release their contents, i.e, soluble mediators and small vesicles, into the intercellular cleft between the T cell and target cell. This mechanism allows a specific and unidirectional delivery of the lytic machinery to the target cell, and may explain why CTLs and bystander cells in close proximity escape killing.[24] However, the way these vesicles interact with the target cell is still poorly understood. It has been proposed that they may fuse with the target cell membrane, or alternàtively they may be endocytosed and deliver their contents to an intracellular compartment of the target cell.

III. CONCLUSIONS AND PERSPECTIVES

Accumulating evidence indicates that cells of the hematopoietic cell lineage are able to secrete the contents of their MVBs, i.e., exosomes and solutes, into the extracellular environment. The subcellular mechanisms involved in exosome biogenesis and release are still poorly understood and future research is needed to yield a model in which exosomes may find their place in cell physiology. Also, it remains to be investigated whether exosome secretion occurs in other nonspecialized secretory cells, since it represents a potential mechanism accounting for so-called nonclassical routes of secretion.

ACKNOWLEDGMENTS

We are grateful to all the members of our laboratories for their contributions to the field and for stimulating discussions. GR is indebted to Dr. D. Louvard (Curie Institut, Paris, France) for his encouragement and support and to Dr G. Langsley (Department of Immunology, Pasteur Institut, Paris, France) for reading the manuscript. R. Schriwaneck (Department of Cell Biology, Utrecht University, the Netherlands) is acknowledged for printing electron micrographs. Finally, we apologize to all the colleagues whose work could not be cited.

REFERENCES

1. Palade G. Intracellular aspects of the process of protein secretion. Science 1975; 189: 347-385.
2. Burgess TL, Kelly RB. Constitutive and regulated secretion of proteins. Ann Rev Cell Biol 1987; 3: 243-93.
3. Pan BT, Teng K, Wu C et al. Electron microscopic evidence for externalization of the transferrin receptor in vesicular form in sheep reticulocytes. J Cell Biol 1985; 101: 942-948.
4. Harding C, Heuser J, Stahl P. Receptor-mediated endocytosis of transferrin and recycling of the transferrin receptor in rat reticulocytes. J Cell Biol 1983; 97: 329-339.
5. Vidal M, Sainte-Marie J, Philippot JR et al. Asymmetric distribution of phospholipids in the membrane of vesicles released during in vitro maturation of guinea pig reticulocytes: evidence precluding a role for "aminophospholipid translocase". J Cell Physiol 1989; 140: 455-62.
6. Raposo G, Nijman HW, Stoorvogel W et al. B lymphocytes secrete antigen-presenting vesicles. J Exp Med 1996 ; 183: 1161-1172.
7. Harding CV, Levy M, Stahl P. Morphological analysis of ligand uptake and processing: the role of multivesicular endosomes and CURL in receptor-ligand processing. Eur J Cell Biol 1985; 36: 230-238.
8. van Deurs B, Holm PK, Kayser L et al. Multivesicular bodies in HEp-2 cells are maturating endosomes. Eur J Cell Biol 1993; 61: 208-224.
9. Trowbridge IS, Collawn JF, Hopkins CR. Signal-dependent membrane protein trafficking in the endocytic pathway. Ann Rev Cell Biol 1993; 9: 129-161.
10. Kornfeld S, Mellman I. The biogenesis of lysosomes. Ann Rev Cell Biol 1989; 5: 483-525.
11. Hunziker W, Geuze HJ. Intracellular trafficking of lysosomal membrane proteins. BioEssays 1996; 18: 379-389.
12. Griffiths GM. Secretory lysosomes-a special mechanism of regulated secretion in haematopoietic cells. Trends Cell Biol 1996; 6: 329-332.
13. Peters PJ, Geuze HJ, van der Donk HA et al. Molecules relevant for T cell-target cell interaction are present in cytolytic granules of human T lymphocytes. Eur J Immunol 1989; 19: 1469-1475.
14. Peters PJ, Borst J, Oorschot V et al. Cytotoxic lymphocyte-T granules are secretory lysosomes, containing both perforin and granzymes. J Exp Med 1991; 173: 1099-1109.
15. Burkhardt JK, Hester S, Lapham CK et al. The lytic granules of natural killer cells are dual-function organelles combining secretory and pre-lysosomal compartments. J Cell Biol 1990; 111: 2327-2340.
16. Dvorak AM. Ultrastructural analysis of human mast cells and basophils. In: Marone G, ed. Human basophils and mast cells. Chem Immunol Basel, Karger: 1995: 1-33. vol 61).
17. Chi EY, Lagunoff D. Abnormal mast cell granules in the beige (Chediak-Higashi syndrome) mouse. J Histochem Cytochem 1975; 23: 117-122.

18. Parmley RT, Poon MC, Crist WM, Malluh A. Giant platelet granules in a child with the Chediak-Higashi syndrome. Am J Hematol 1979; 6: 51-60.
19. White JG, Clawson CC. The Chediak-Higashi syndrome; the nature of the giant neutrophil granules and their interactions with cytoplasm and foreign particulates. Progressive enlargement of the massive inclusions in mature neutrophils. Am J Pathol 1980; 98: 151-196.
20. Harding C, Heuser J, Stahl P. Endocytosis and intracellular processing of transferrin and colloidal-gold transferrin in rat reticulocytes: demonstration of a pathway for receptor shedding. Eur J Cell Biol 1984; 35: 256-263.
21. Johnstone RM, Adam M, Hammond JR et al. Vesicle formation during reticulocyte maturation. Association of plasma membrane activities with released vesicles (exosomes). J Biol Chem 1987; 262: 9412-9420.
22. Vidal M, Stahl PD. The small GTP-binding proteins Rab 4 and ARF are associated with released exosomes during reticulocyte maturation. Eur J Cell Biol 1993; 60: 261-267.
23. Griffiths GM. The cell biology of CTL killing. Curr Opin Immunol 1995; 7: 343-348.
24. Peters P, Geuze HJ, van der Donk HA et al. A new model for lethal hit delivery by cytotoxic T lymphocytes. Immunol Today 1990; 11: 28-32.
25. Gruenberg J, Howell KE. Membrane traffic in endocytosis: insights from cell-free systems. Ann Rev Cell Biol 1989; 5: 453-481.
26. Hopkins CR, Trowbridge IS. Internalization and processing of transferrin and transferrin receptors in human carcinoma cells. J Cell Biol 1983; 97: 508-521.
27. Geuze HJ, Slot JW, Strous GJ et al. Intracellular site of asialoglycoprotein receptor-ligand uncoupling: double label immunoelectron microscopy during receptor-mediated endocytosis. Cell 1983; 32: 277-287.
28. Geuze HJ, Slot JW, Schwartz AL. Membranes of sorting organelles display lateral heterogeneity in receptor distribution. J Cell Biol 1987; 104: 1715-1723.
29. Tooze J, Hollinshead M. Tubular early endosomal networks in AtT20 and other cells. J Cell Biol 1991; 115: 635-653.
30. van Deurs B, Petersen OW, Olsnes S et al. The ways of endocytosis. Int Rev Cytol 1989; 117: 131-177.
31. Courtoy PJ. Dissection of endosomes. In: JC Steer JH, ed. Trafficking of Membrane Proteins. New York: 1991: 103-156.
32. Goldstein JL, Brown MS, Anderson RGW et al. Receptor-mediated endocytosis: concepts emerging from the LDL receptor system. Ann Rev Cell Biol 1985; 1: 1-39.
33. Stoorvogel W, Geuze HJ, Strous GJ. Sorting of endocytosed transferrin and asialoglycoprotein occurs immediately after internalization in HepG2 cells. J Cell Biol 1987; 104: 1261-1268.
34. Stoorvogel W, Geuze HJ, Griffith JM et al. Relations between the intra-

cellular pathways of the receptors for transferrin, asialoglycoprotein, and mannose 6-phosphate in human hepatoma cells. J Cell Biol 1989; 108: 2137-2148.

35. Dunn KW, McGraw TE, Maxfield FR. Interactive fractionation of recycling receptors from lysosomally destined ligands in an early sorting endosome. J Cell Biol 1989; 109: 3303-3014.

36. Hopkins CR, Gibson A, Shipman M et al. In migrating fibroblasts, recycling receptors are concentrated in narrow tubules in the pericentriolar area, and then routed to the plasma membrane of the leading lamella. J Cell Biol 1994; 125: 1265-1274.

37. Gosh RN, Maxfield FR. Evidence for nonvectorial, retrograde transferrin trafficking in the early endosomes of HEp2 cells. J Cell Biol 1995; 128: 549-561.

38. Stoorvogel W, Oorschot V, Geuze HJ. A novel class of clathrin-coated vesicles budding from endosomes. J Cell Biol 1996; 132: 21-33.

39. Griffiths G, Gruenberg J. The arguments of pre-existing early and late endosomal compartments. Trends Cell Biol 1991; 1: 5-9.

40. Roederer M, Barry JR, Wilson RB et al. Endosomes can undergo an ATP-dependent density increase in the absence of dense lysosomes. Eur J Cell Biol 1990; 51: 229-234.

41. Stoorvogel W, Strous GJ, Geuze HJ et al. Late endosomes derive from early endosomes by maturation. Cell 1991; 65: 417-427.

42. Murphy RF. Maturation models for endosome and lysosome biogenesis. Trends Cell Biol 1991; 1: 77-82.

43. Futter CE, Pearse A, Hewlett LJ et al. Multivesicular endosomes containing internalized EGF-EGF receptor complexes mature and then fuse directly with lysosomes. J Cell Biol 1996; 132: 1011-1023.

44. Hopkins CR, Gibson A, Shipman M, Miller K. Movement of internalized ligand-receptor complexes along a continuous endosomal reticulum. Nature 1990; 346: 335-339.

45. Felder S, Miller K, Moehren G, Ullrich A, Schlessinger J, Hopkins CR. Kinase activity controls the sorting of the epidermal growth factor receptor within the multivesicular body. Cell 1990; 61: 623-634.

46. Futter CES, Felder S, Schlessinger J et al. Annexin 1 is phosphorylated in the multivesicular body during the processing of the epidermal growth factor. J Cell Biol 1993; 120: 77-83.

47. Schlossman DM, Schmid SL, Braell WA et al. An enzyme that removes clathrin coats: purification of an uncoating ATPase. J Cell Biol 1984; 99: 723-733.

48. Davis JQ, Dansereau D, Johnstone RM et al. Selective externalization of an ATP-binding protein structurally related to the clathrin-uncoating ATPase/heat shock protein in vesicles containing terminal transferrin receptors during reticulocyte maturation. J Biol Chem 1986; 261: 15368-15371.

49. Kurten RC, Cadena DL, Gill GN. Enhanced degradation of EGF receptors by a sorting nexin, SNX1. Science 1996; 272: 1008-1010.

50. Geuze HJ, Stoorvogel W, Strous GJ et al. Sorting of mannose-6-phosphate receptors and lysosomal membrane proteins in endocytic vesicles. 107 1988; 2491-2501.
51. Klumperman J, Hille A, Veenendaal T et al. Differences in the endosomal distributions of the two mannose-6-phosphate receptors. J Cell Biol 1993; 121: 997-1010.
52. Pfeffer SR, Rothman JE. Biosynthetic protein transport and sorting by the endoplasmic reticulum and Golgi. Ann Rev Biochem 1987; 56: 829-852.
53. Matter K, Mellman I. Mechanisms of cell polarity: sorting and transport in epithelial cells. Curr Opin Cell Biol 1994; 6: 545-554.
54. Simons K. Biogenesis of epithelial cell surface polarity. Harvey Lect. 1995; 89: 125-146.
55. Yoshimori T, Keller P, Roth MG et al. Different biosynthetic transport routes to the plasma membrane in BHK and CHO cells. J Cell Biol 1996; 133: 247-256.
56. Gerdes HH, Rosa P, Phillips E et al. The primary structure of hman secretogranin II: a widespread tyrosine-sulfated secretory protein that exhibits low pH and a calcium induced aggregation. J Biol Chem 1989; 264: 12009-12015.
57. Wagner DD, Saffaripour S, Bonfanti R et al. Induction of specific storage organelles by von Willebrand factor propolypeptide. Cell 1991; 64: 403-413.
58. Chanat E, Huttner WB. Milieu-induced, selective aggregation of regulated secretory proteins in the trans-Golgi network. J Cell Biol 1991; 115: 1505-1519.
59. Bauerfeind R, Huttner WB. Biogenesis of constitutive secretory vesicles, secretory granules and synaptic vesicles. Curr Opin Cell Biol 1993; 5: 628-635.
60. Kuliawat R, Arvan P. Distinct molecular mechanisms for protein sorting within immature secretory granules of pancreatic β-cells. J Cell Biol 1994; 126: 77-86.
61. Tapper H. The secretion of preformed granules by macrophages and neutrophils. J Leukoc Biol 1996; 59: 613-622.
62. Voorhout WF, Weaver TE, Haagsman HP et al. Biosynthetic routing of pulmonary surfactant proteins in alveolar type II cells. Microsc Res Tech 1993; 26: 366-373.
63. Baron R. Molecular mechanisms of bone resorption. An update. Acta Orthopaedica Scandinavica Suppl. 1995; 266: 66-70.
64. Podack ER, Young JDE, Cohn ZA. Isolation and characterization of perforin 1 from cytolytic T cell granules. Proc Natl Acad Sci USA 1985; 82: 8629-8633.
65. Schmid SL, Rothman JE. Enzymatic dissociation of clathrin cages in a two-stage process. J Biol Chem 1985; 260: 10044-10049.
66. Bu G, Schwartz AL. Receptor-mediated endocytosis. The liver: Biology and pathology. Third ed. New-York: Raven Press, Ltd., 1994; 259-274.

67. Li CY, Watkins JA, Glass J. The $^+$H-ATPase from reticulocyte endosomes reconstituted into liposomes acts as an iron transporter. J Biol Chem 1994; 269: 10242-10246.

68. Dautry-Varsat A, Ciechanover A, Lodish HF. pH and the recycling of transferrin during receptor-mediated endocytosis. Proc Natl Acad Sci USA 1983; 80: 2258-2262.

69. van Bockxmeer FM, Morgan EH. Transferrin receptors during rabbit reticulocyte maturation. Bioch Biophys Acta 1979; 584: 76-83.

70. Johnstone RM. The transferrin receptor. In: Parker PAaJC, ed. Hematology: red blood cell membranes. New York: M. Dekker, 1989; 11: 325-365.

71. Johnson LS, Dunn KW, Pytowski B et al. Endosome acidification and receptor trafficking: bafilomycin A1 slows receptor externalization by a mechanism involving the receptor's internalization motif. Mol Biol Cell 1993; 4: 1251-1266.

72. Aniento F, Parton RG, Gruenberg J. An endosomal beta COP is involved in the pH-dependent formation of transport vesicles destined for late endosomes. J Cell Biol 1996; 133: 29-41.

73. Simons K, Wandinger-Ness A. Polarized sorting in epithelia. Cell 1990; 62: 207-210.

74. Sandhoff K, Klein A. Intracellular trafficking of glycosphingolipids: role of sphingolipid activator proteins in the topology of endocytosis and lysosomal digestion. FEBS Lett 1994; 346: 103-107.

75. Felder S, Miller K, Moehren G et al. Kinase activity controls the sorting of the epidermal growth factor receptor within the multivesicular body. Cell 1990; 61: 623-634.

76. Shih YJ, Baynes RD, Hudson BG et al. Serum transferrin is a truncated form of the tissue receptor. J Biol Chem 1990; 265:19077-19081.

77. Cresswell P. Assembly, transport, and function of MHC class II molecules. Ann Rev Immunol 1994; 12: 259-293.

78. Wolf PR, Ploegh HL. How MHC class II molecules acquire peptide cargo: Biosynthesis and trafficking through the endocytic pathway. Ann Rev Cell Dev Biol 1995; 11: 267-306.

79. Marks MS, Blum JS, Cresswell P. Invariant chain trimers are sequestered in the RER in the absence of association with HLA class II antigens. J Cell Biol 1990; 111: 839-855.

80. Cresswell P. Invariant chain structure and MHC class II function. Cell 1996; 84: 505-507.

81. Neefjes JJ, Stollorz V, Peters PJ et al. The biosynthetic pathway of MHC class II but not class I molecules intersects the endocytic route. Cell 1990; 61: 171-183.

82. Peters PJ, Neefjes JJ, Oorschot V et al. Segregation of MHC class-II molecules from MHC class I molecules in the Golgi complex for transport to lysosomal compartments. Nature 1991; 349: 669-676.

83. Morton PA, Zacheis ML, Giacoletto KS et al. Delivery of nascent MHC

class II-invariant chain by cystein proteases precedes peptide binding in B-lymphoblastoid cells. J Immunol 1995; 154: 137-150.

84. Riese RJ, Wolf PR, Brömme D et al. Essential role for cathepsin S in MHC class II-associated invariant chain processing and peptide loading. Immunity 1996; 4: 357-366.

85. Sloan VS, Camerson P, Porter G et al. Mediation by HLA-DM of dissociation of peptides from HLA-DR. Nature 1995; 375: 802-806.

86. Denzin LK, Cresswell P. HLA-DM induces CLIP dissociation from MHC class II α/β dimers and facilitates peptide loading. Cell 1995; 82: 155-165.

87. Germain RN. MHC-dependent antigen processing and peptide presentation: providing ligands for T lymphocyte activation. Cell 1994; 76: 287-299.

88. Kleijmeer M, Oorschot V, Geuze HJ. Human Langerhans cells display a lysosomal compartment enriched in MHC class II. J Invest Dermatol 1994; 103: 516-523.

89. Nijman HW, Kleijmeer MJ, Ossevort MA et al. Antigen capture and MHC class II compartments in freshly isolated and cultured blood dendritic cells. J Exp Med 1995; 182: 163-174.

90. Peters P, Raposo G, Neefjes JJ et al. MHC class II compartments in human B lymphoblastoid cells are distinct from early endosomes. J Exp Med 1995; 325-334.

91. Glickman JN, Morton PA, Slot JW et al. The biogenesis of MHC class II compartment in human I-cell disease B lymphoblasts. J Cell Biol 1996; 132: 769-785.

92. Qiu Y, Xu X, Wandinger-Ness A et al. Separation of subcellular compartments containing distinct functional forms of MHC class II. J Cell Biol 1994; 125: 595-605.

93. Rudensky AY, Maric M, Eastman S et al. Intracellular assembly and transport of endogenous peptide-MHC class II complexes. Immunity 1994; 1: 585-594.

94. Tulp A, Verwoerd D, Dobberstein B et al. Isolation and characterization of the intracellular MHC class II compartment. Nature 1994; 369: 120-126.

95. West MA, Lucocq JM, Watts C. Antigen processing and class II MHC peptide-loading compartments in human B-lymphoblastoid cells. Nature 1994; 369: 147-151.

96. Sanderson F, Kleijmeer MJ, Kelly A et al. Accumulation of HLA-DM, a regulator of antigen presentation, in MHC class II compartments. Science 1994; 266: 1566-1569.

97. Amigorena S, Drake JR, Webster P et al. Transient accumulation of new class II MHC molecules in a novel endocytic compartment in B lymphocytes. Nature 1994; 369: 113-120.

98. Castellino F, Germain RN. Extensive trafficking of MHC class II-invariant chain complexes in the endocytic pathway and appearance of peptide-loaded class II in multiple compartments. Immunity 1995; 2: 73-88.

99. Lippincott-Schwartz J, Fambrough DM. Cycling of the integral membrane glycoprotein, LEP 100, between plasma membrane and lysosomes: kinetic and morphological analysis. Cell 1987; 49: 669-677.

100. Harter C, Mellman I. transport of the lysosomal membrane glycoprotein lgp 120 (lgp-A) to lysosomes does not require appearance on the plasma membrane. J Cell Biol 1992; 117: 311-325.

101. Emerson SG, Cone RE. Turnover and shedding of Ia antigens by murine spleen cells in culture. J Immunol 1979; 122: 892-899.

102. Emerson SG, Cone RE. I-Kk and H-2Kk antigens are shed as supramolecular particles in association with membrane lipids. J Immunol 1981; 127: 482-486.

103. Sachs DH, Kiszkiss P, Kim KJ. Release of Ia antigens by a cultured B cell line. J Immunol 1980; 124: 2130-2136.

104. Callahan GN, Ferrone S, Poulik MD et al. Characterization of Ia antigens in mouse serum. J Immunol 1976; 117: 1351-1355.

105. Gray D, Kosco M, Stockinger B. Novel pathways of antigen presentation for the maintenance of memory. Int Immunol 1991; 3: 141-148.

106. Meeusen E. The induction of cytotoxic T cell responses with H-2 antigens shed from viable lymphocytes. Immunol 1987; 61: 321-326.

107. Masson D, Tschopp J. Isolation of a lytic, pore-forming protein (perforin) from cytolytic T lymphocytes. J Biol Chem 1985; 260: 9069-9072.

108. Baetz K, Isaaz S, Griffiths GM. Loss of cytotoxic T lymphocyte function in Chediak-Higashi syndrome arises from a secretory defect that prevents lytic granule exocytosis. J Immunol 1995; 154: 6122-6131.

109. Young JD, Liu CC, Persechini PM, Cohn ZA. Perforin-dependent and -independent pathways of cytotoxicity mediated by lymphocytes. Immunological Rev 1988; 103: 161-202.

INDEX